W9-BXR-045

Comparative Physiology *of* Respiratory Mechanisms

Comparative Physiology *of* Respiratory Mechanisms

JOHAN B. STEEN

Institute of Physiology
University of Oslo, Norway

1971

ACADEMIC PRESS · LONDON · NEW YORK

ACADEMIC PRESS INC. (LONDON) LTD
Berkeley Square House,
Berkeley Square,
London, W1X 6BA

U.S. Edition published by
ACADEMIC PRESS INC.
111 Fifth Avenue,
New York, New York 10003

Library of Congress Catalog Card Number: 70-129796

ISBN: 0·12-664150-1

Printed in Great Britain by
Alden & Mowbray Ltd
at the Alden Press, Oxford

PREFACE

This book is an attempt to present what I think are the essential features of the respiratory mechanisms of animals. I have tried to summarize current knowledge on the various subjects and to indicate problems of future research. No attempt has been made to give a complete review of the existing literature.

The central issue of the book is gas exchange between the animal and the environment. Related issues, like regulation of respiration, and maintenance of an appropriate internal composition have only been touched upon. But since the respiratory function is integrated with the other functions of the organisms, I have also considered how the respiratory function fits the organism as a whole.

Despite these limitations the subject is huge, so huge that I have probably not been able to discuss all aspects of it with the competence required by specialists in each field.

The enormous number of publications dealing with comparative respiratory physiology prohibits detailed discussion and reference to all but a few original papers. Many more have been consulted than appear in the list of references and doubtless a number has been overlooked.

The title suggests that this book is a new version of August Krogh's classical little book published 30 years ago. This is so in the sense that it takes up many of the same subjects. I am afraid, however, that I have not been able to attain his simple and to-the-point manner of discussion. Krogh's clear mind made it unnecessary for him to use anything but the simplest language. I have not been able to achieve this although I have tried.

During the preparation of this book I have benefited from the advice and criticism from a number of collegues. I am particularly thankful to (listed in alphabetical order): Drs. H. T. Andersen, T. Berg, P. S. Enger, K. Johansen, J. Krogh, H. Kutchai, G. Nicolaysen, K. Schmidt-Nielsen and B. A. Waaler. Their advice has greatly improved several aspects of the book, but they are not responsible for the errors which remain.

I will also use this opportunity to acknowledge the co-operation of Dr. I. Steen who helped me to review the literature. Mrs. Eva Fosse has typed and retyped the manuscript with an optimum of charm, skill and patience.

CONTENTS

Preface v

Chapter 1
PRINCIPAL PROPERTIES OF GAS EXCHANGE

A. Diffusion 1
B. The Performance of Respiratory Mechanisms 5
C. Oxygen Transport by Blood 7
D. Carbon Dioxide Transport of Blood 11
E. The Rate of Reactions of Respiratory Gases with Blood . . . 13

Chapter 2
THE RESPIRATORY ENVIRONMENTS

A. Air 16
B. Water 16
C. Natural Waters 17
D. Biological Consequences of the Difference Between Air and Water as Respiratory Media 19

Chapter 3
AQUATIC GAS EXCHANGE

A. Gas Exchange Without Respiratory Organs, Circulatory Systems or Respiratory Pigments 21
B. Gas Exchange with Primitive Respiratory Organs and Circulation, but Without Pigment 22
C. Gas Exchange with a Respiratory Organ, Organized Ventilation and Respiratory Pigment, but an Open Circulatory System 25
D. Gas Exchange with Specialized Respiratory System, Organized Ventilation, Closed Circulatory System and a Respiratory Pigment in Blood Cells 34
   1. The Structure of the Gill Apparatus 34
   2. Functional Analysis of the Gill Apparatus 37

Chapter 4
TRANSITIONAL BREATHING

A. The Transition from Aquatic to Aerial Breathing 56
   1. Reactions to Hypoxic Water 57
   2. Invertebrate Dual Breathers 57
   3. Air-breathing Fishes 58
B. Respiratory Adaptations Accompanying the Amphibian Metamorphosis 75

Chapter 5
AERIAL GAS EXCHANGE

A. Gas Exchange Across the General Body Surface with Circulation and a Respiratory Pigment 77
B. Gas Exchange by Tubes which lead Directly to the Tissues . . . 80
1. The Architecture of the Tracheal System 80
2. Gas Renewal of the Tracheal System 82
3. Gas Exchange During Flight 83
4. Comparison Between the Tracheal System and the Vertebrate Capillary System 86
5. Gas Exchange and Water Economy—Cyclic CO_2-Release . . 87
6. Adaptation of the Tracheal System of Aquatic Life—Plastron Breathing 92
C. Respiration by Lungs and a Circulatory System with Respiratory Pigment 95
1. Diffusion Lungs 95
2. Ventilation Lungs—General Properties 98
D. The Alveolar Lung 100
1. Structure 100
2. Stability of the Alveoli 102
3. Ventilation and Renewal of Gas 104
4. Regulation of Circulation and Ventilation 104
5. Respiratory Properties Related to the Shape and Size of the Body . 105
E. The Air Capillary Lung 114
1. Organization of the Airways 115
2. Gas Exchange Within the Respiratory Unit 116
3. Ventilation 117
4. Distribution of Gas 118

Chapter 6
RESPIRATORY ADAPTATIONS TO LIFE AT HIGH ALTITUDE

124

Chapter 7
SECONDARY ADAPTATIONS OF LUNG BREATHERS TO AQUATIC LIFE

A. Accessory Water Breathing in Vertebrates 132
B. Adaptations to Prolonged Submersion 132
1. Blood Properties and O_2-Exchange During Submersion . . . 133
2. Defence Against Decompression Sickness 135
3. Cardiovascular Adaptations 135
4. Energy Production 136
5. Chemical Sensitivity of Ventilation 138

Chapter 8
PLACENTAL GAS EXCHANGE

A. Structure 139
B. Gas Exchange 140
C. Blood flow 141
D. Blood 141

Chapter 9
GAS EXCHANGE OF THE BIRD EGG 146

Chapter 10
THE SWIMBLADDER FUNCTION 155

References 162
Author Index 173
Subject Index 177

Chapter I

PRINCIPAL PROPERTIES OF GAS EXCHANGE

The extent of gas exchange, i.e. the O_2-uptake or CO_2-output of an organism, is determined by its metabolic demands, which depend on a number of factors like activity and size. We shall not be concerned with these aspects of respiration, but rather with where and how this gas exchange occurs.

Gases exchange across respiratory membranes by diffusion. Before we begin the discussion of the respiratory mechanisms in the animal kingdom, we shall therefore briefly review some pertinent aspects of the basic diffusion process.

A. Diffusion

The process of diffusion reflects a basic phenomenon of nature, namely that matter will flow spontaneously from areas of high to areas of low free energy. In the case of gases, free energy is proportional to partial pressure (P). A pure gas has a partial pressure equal to its pressure as measured by, for example, a manometer. In a gas mixture each component will exert a partial pressure equal to the total pressure times the volume proportion of the component. Air contains 20·93% O_2, 0·03% CO_2 and 79·04% inert gases. If the total pressure is 1 atmosphere, then the partial pressure of O_2 (P_{O_2}) is 0·2093 atm, the P_{CO_2} is 0·0003 atm and the $P_{\text{inert gases}} = 0{\cdot}7904$ atm. Since 1 atm equals 760 mm Hg the $P_{O_2} = 159{\cdot}1$ mm Hg, $P_{CO_2} = 0{\cdot}2$ mm Hg and $P_{\text{inert gases}} = 600{\cdot}7$ mm Hg.

If this air sample is heated at constant volume until the pressure is doubled, the partial pressures will also be doubled. However, if it is heated at constant pressure, the partial pressures will remain constant; this is because the partial pressure is not proportional to the concentration of the gas molecules alone, but rather to the concentration times the average free energy of the molecules. This is expressed in the gas law as $PV = nRT$ where P is the pressure, V the volume, T the temperature, n the number of moles of the gas and R the gas constant. The product RT is proportional to the free energy of 1 mole of gas. It is useful to recall that 1 mole of a gas ($6{\cdot}023 \times 10^{23}$ molecules) at 1 atm will occupy 22·4 litre at 0°C. If one mole of gas is compressed to 1 litre it exerts a pressure of 22·4 atm at 0°C.

If a mixture of gases, for example air, is enclosed in a container together with water, part of the gases will dissolve in water. Only CO_2 reacts chemically with water, the others simply take up places between the molecules of water or aggregates of these. They are physically dissolved.

When equilibrium is established in such a water–gas system the free energy of the gases, and hence their partial pressures, is identical in the two phases. The partial pressure of a gas dissolved in water is therefore equal to, and often measured as, its partial pressure in a gas phase with which the water is in equilibrium. The partial pressure of a gas dissolved in water is often termed the gas *tension*.

Each gas will dissolve in the water independently of the others and in proportion to its partial pressure. At equilibrium the concentration (C) of each gas dissolved in water is equal to the product of its partial pressure (P) and its solubility coefficient (α), $C = \mathrm{P}\alpha$.

The solubility coefficient of a gas expresses the ability of water to accommodate the gas molecules. It is defined as the volume of dry gas (at 0°C and one atm) contained in one volume of fluid at a partial gas pressure of 760 mm Hg. For respiratory purposes it is conveniently expressed as the volume gas (STPD) dissolved in 100 volumes of water (volume per cent).

If two gases, two fluids, or a fluid and a gas, contain a particular gas at different partial pressures, gas exchange will occur as long as the two are not separated by an impermeable barrier. Gas will move from the phase where its partial pressure is higher to that where it is lower; and since the metabolism of living non-photosynthesizing organisms continuously removes O_2 and produces CO_2, there will always be a difference in the partial pressure of the respiratory gases between the inside and the outside of such organisms. Since no biological material is impermeable to gases, diffusion will always take place between an organism and its environment. In higher animals such exchange is localized in the respiratory organs.

Until early in this century there was serious, and at the time apparently well founded, doubt that gas exchange by diffusion could account for the entire gas exchange in the lungs. August Krogh (1910) developed, most typically for the great experimentalist, methods that could measure the P_{O_2} and P_{CO_2} of alveolar gas and arterial blood sufficiently accurately to settle the argument. Physiologists before him, among them his teacher and close friend Christian Bohr, had employed inadequate methods and found higher P_{O_2} in arterial blood than in alveolar gas. This, they correctly argued, could not be due to diffusion. Bohr therefore postulated that gases were secreted into the blood. Krogh was unable to confirm these results. The paper (Krogh, 1910) where he describes these findings and points out the errors of his professor, is worth reading.

Today diffusion is widely accepted as the only mechanism of gas exchange. This is partly due to Krogh's measurements, but mostly to the fact that no case of active transport of O_2 or CO_2 (secretion) across a biological membrane has been demonstrated. The mechanism of gas concentration in the swimbladder is no exception (Ch. 10).

In the respiratory organs the internal fluid and the environment are separated by a large, thin and gas-permeable membrane.

The amount Q of gas diffusing across a membrane of thickness L and area A per unit time is equal to:

$$Q = D\frac{A\Delta P}{L} \quad (1)$$

where ΔP is the difference in gas partial pressure across the membrane and D a coefficient expressing the permeability of the membrane to the gas in question.

These variables are of different nature. The amount of gas exchange, O_2-uptake or CO_2-output, reflects the metabolic needs of the animal and its energy requirements. This gas exchange takes place across some barrier. The dimensions of this barrier are given by its area and thickness. The gas permeability is based on certain physical and chemical properties of the membrane and the gas, while the partial pressure difference depends on the nature, composition and renewal rate of external and internal environments. Thus, three variables in equation (1) express properties of the membrane and one the force causing gas diffusion across it.

We shall, in the following, focus interest on some determinants of these variables which are important for a quantitative analysis of a respiratory organ.

Surface area and thickness. These two parameters are primarily determined by the anatomy of the respiratory organ. The external area, i.e. that facing the environment, can be measured fairly accurately in organs where the respiratory surface is smooth or folded in a regular pattern. In organs where the surface has an irregular geometry the area can be obtained only after extensive measurements and mathematical treatment of the data. The surface area of the membrane facing the body fluid is much more difficult to obtain, particularly in animals where blood flows in capillaries. The thickness of the respiratory membrane can be obtained by histological techniques. Respiratory membranes are, however, frequently so thin that their thickness cannot be measured in the light microscope.

Such structural values only rarely represent the effective values in the functioning organ. This is due, first of all, to incomplete renewal of internal and external medium at the two sides of the measured membrane. Part of the organ may not be circulated and/or ventilated, which reduces the effective diffusion area. The linear velocity of fluid flow at the surface will determine the thickness of the unstirred layer. This adds a fluid layer to the tissue layer and increases the diffusion path. All these factors tend to make the functioning surface area smaller and the diffusion path longer than the values obtained from anatomical and histological measurements.

The partial pressure difference. In a simple experimental situation where a membrane separates two reservoirs, each containing well-stirred water with constant P_{O_2}, the ΔP_{O_2} is easily obtained as the difference between the P_{O_2} of the two compartments. Such a simple situation unfortunately never exists in a respiratory organ. Except for the most primitive animals both external and internal medium is in contact with the membrane only for a short time, during which the P_{O_2} either increases or decreases. In the case where both media contain only dissolved O_2 the mean P_{O_2} on either side can be calculated from the linear flow velocity and the P_{O_2} of the medium before and after it

TABLE 1.1

Permeability coefficients (from Krogh, 1920).

Material	Permeability coefficients at 20°C (ml gas(STP)/cm²-min-atm/μ)
Air	110,000
Water	0·34
Gelatine	0·28
Muscle	0·14
Connective tissue	0·11
Chitin	0·13

has passed the membrane. Such P_{O_2} values may be difficult to obtain from a functioning respiratory organ. Frequently the fluid or gas that has actually been in contact with the respiratory surface is mixed with fluid or gas that has not, or to a lesser extent, been in exchange contact with the surface. The P_{O_2} of samples of expired gas or water may therefore not be representative of the P_{O_2} at the respiratory surface. The situation is further complicated when the blood contains a respiratory pigment which causes a non-linear relationship between P_{gas} and gas content.

Permeability. The permeability of a respiratory membrane is a manifestation of certain of its physico-chemical properties. Our insight is, however, far from sufficient to predict the permeability of a tissue from its composition. Instead we must measure the other variables of the diffusion equation and calculate the permeability. This is expressed as a permeability coefficient D which expresses the amount of gas in millilitres diffusing per minute through 1 cm^2 area and 1 μ thickness at a partial pressure difference of 1 atm (760 mm Hg). Other units may of course be used for D, but these are the most convenient for our purpose. Krogh (1920) measured the permeability of O_2 through non-respiring tissue slices and other materials (Table 1.1) and thus not under conditions comparable to those existing in a living animal. However, since permeability values of intact, respiring and normally perfused tissues are difficult to obtain, Krogh's 50-year-old values are still widely used.

As can be seen from Table 1.1, all Krogh's permeabilities for tissues were slightly below the permeability of O_2 in water. This fact together with later measurements has led to the generalization that the permeability of a tissue is approximately equal to that of water times the fraction of the tissue composed of water. While this may be correct for the diffusion process itself the exchange process may be augmented in several ways.

One possibility is that gas exchange is facilitated by intracellular convection. Such convection is a normal property of living cells (protoplasmic streaming). There is some indirect experimental evidence to support this view. Longmuir and Bourke (1960) and MacDougall and McCobe (1967) calculated D from measurements of O_2-uptake by tissue slices. Their values are from 2 to 10 times higher than D for water. It is known that the intracellular fluid streaming is very sensitive to the O_2-supply.

It is also interesting that slight mechanical vibration may increase the apparent value for D drastically (Longmuir and Bourke, 1960). This suggests that "physiological vibration" like the rhythmic pumping of the heart, the movements of lung ventilation, variation in muscle tonus or a massage effect of red corpuscle on the capillary walls, may possibly facilitate gas exchange *in situ*. Such possibilities have not yet been examined.

Another possibility of enhanced O_2-exchange was discovered by Scholander (1960) who showed that the presence of Hb in a diffusion barrier can facilitate the rate of O_2-flux as much as 8-fold. The mechanism of this facilitation has been disputed (Hemmingsen, 1965). Scholander's original idea was a bucket-brigade model where O_2-molecules were "handed" from one Hb molecule to the next. Another explanation is that facilitation is accomplished by diffusion of Hb molecules (Moll, 1966). It is not clear if, and to what extent, this facilitation contributes to the rate of O_2-exchange in the intact animal. In this connection it is worth considering that neither the lung membrane nor the gill barrier of fishes contains intracellular Hb.

B. The Performance of Respiratory Mechanisms

We have so far considered how the diffusion parameters are involved in the process of gas exchange. In order to compare gas-exchange processes of various organisms we must assess the performance of a respiratory system in quantitative terms.

The *exchange* properties of the organ itself may be expressed by a diffusion capacity that is the ratio of the amount of gas exchanged to the difference in partial pressure across the respiratory membrane. To be useful for comparative purposes, the diffusion capacity must be expressed per unit area. This reduces the diffusion capacity to the permeability divided by the thickness of the barrier (see equation (1)). Partly because the permeability has been little investigated, it is often assumed to be the same for all respiratory membranes. If this is true, then diffusion capacity and thus the exchange properties of a respiratory organ may, for practical purposes, be expressed by the thickness of the barrier. We can thus draw the self-evident conclusion that a respiratory organ with a thin barrier between internal and external medium has a high diffusion capacity.

The most meaningful measures of the *performance* of a process are its capacity and in particular its efficiency. The capacity of a respiratory mechanism is simply the maximum O_2-uptake of the animal. This parameter is, however, not particularly descriptive for the respiratory mechanism since it depends upon the size of the animal.

The respiratory efficiency ought to be the best measure of the performance of a respiratory system. It should ideally express the ratio of gas exchange to the energy expended to bring about this exchange. This energy is mostly that required for ventilation and circulation. However, these quantities are very difficult to measure, so difficult that it has been attempted only in a few cases and with limited success. However, if we accept the rather questionable assumption that the energy of ventilation and circulation is proportional to the flow and independent of the propulsive mechanism, we can use the O_2-extraction from the environment and the arterio-venous difference in O_2-content as an approximate measure of efficiency.

The pattern of ventilation and circulation is an important factor in determining the efficiency, i.e. gas extraction of a respiratory organ. One aspect of this pattern is separation of arterial from venous blood and of inhaled from exhaled gas or water. To appreciate the importance of separation of input and output fluid we shall recall that, other factors being constant, the gas exchange is proportional to the partial pressure difference across the barrier. If, for example, well-oxygenated blood from the respiratory organ is mixed with deoxygenated blood before it enters the tissues, then its P_{O_2} will be reduced and the ΔP_{O_2} between blood and tissue will also be reduced. Such conditions exist in lower animals and must reduce the efficiency of their respiratory systems.

The flow pattern across the respiratory surface is also of importance. We can distinguish between two extremes: con-current and counter-current flow. These are illustrated in Fig. 1.1. Let the input concentration be 0 and 100 and the flow be equal on both sides of the common membrane. Then the output concentration in the con-current system will approach 50 while those in the counter-current system will approach 100 and 0. The counter-current pattern therefore provides a more efficient system than the con-current. Some of the more sophisticated aspects of counter-current exchange are discussed in Chapter 10.

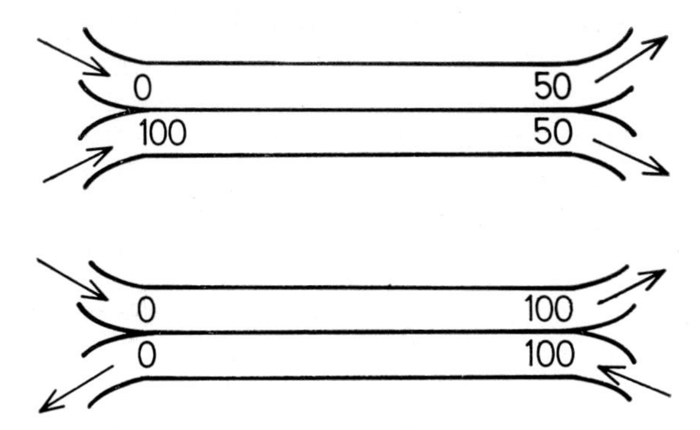

Fig. 1.1. Schematic representation of an exchange system with con-current flow (top) and counter-current flow (bottom). The flow is the same in all tubes and the numbers represent P_{O_2}. The outflow P_{O_2} are for the limiting condition, i.e. maximum exchange.

C. Oxygen Transport by Blood

The primitive type of blood or haemolymph which contains no respiratory pigments, has the same respiratory properties as salt water (Ch. 2). The presence of respiratory pigments gives the blood important respiratory qualities. All respiratory pigments have certain properties in common, both chemical and functional. They all contain a protein and a heavy metal ion. Haemoglobin (Hb) is composed of the protein globin and the prosthetic group haeme with a ferrous ion (Fe^{2+}) in the centre. Recent X-ray diffraction studies have revealed the detailed structure of this intriguing molecule (Perutz, 1969).

Most haemoglobins are built of units of molecular weight 16–17 000. Each subunit contains one haeme, one Fe^{2+} and one globin. All vertebrates except cyclostomes have haemoglobins with a molecular weight of about 68 000 and hence consist of 4 subunits. Each subunit can combine with 1 molecule of O_2, i.e. there can be a maximum of 1 O_2-molecule per Fe^{2+} or per 16–17 000 atomic mass units. A vertebrate Hb therefore has an O_2-capacity of 4 mol O_2 per mol Hb or 1·34 ml O_2 per g Hb.

Haemocyanin (Hcy) contains copper and can usually combine with 1 mol O_2 per 50 000–75 000 g of Hcy. In most cases, however, Hcy occurs in larger complexes.

The most important respiratory property of these pigments is the ability to combine reversibly with O_2 at partial pressures below 150 mm Hg. The presence of such pigments therefore gives the blood (1) an increased content of respiratory gases and (2) the ability to act as a carrier loading and unloading gas at the appropriate areas.

The relationship between the P_{O_2} and the amount of O_2 bound to blood, or percentage O_2-saturation of the pigment, is conveniently illustrated in O_2-equilibrium curves (Fig. 1.2). The most informative of such curves are those where the vertical axes express O_2-content rather than percentage saturation. In this case the curve gives information both about O_2-capacity and about the pressure range where O_2 can be reversibly bound. If blood

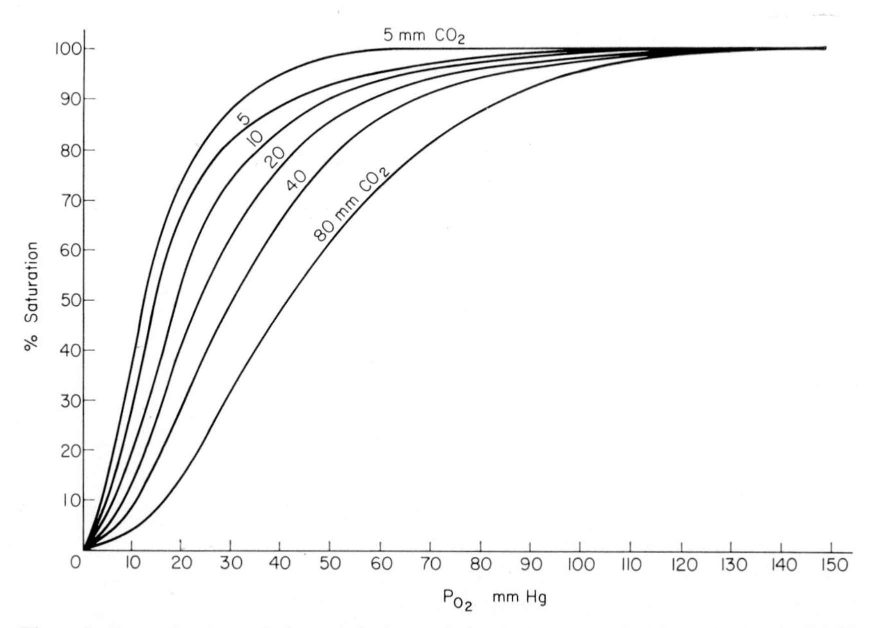

Fig. 1.2. Reproduction of the original graph by Bohr, Hasselbalch and Krogh (1904) demonstrating the effect of P_{CO_2} upon the O_2-affinity of human blood (the Bohr-effect). Ordinate percentage O_2-saturation; abscissa P_{O_2}, temperature: 37°C.

were water, it could carry only about 0·2 ml of O_2 in each hundred ml. Vertebrate bloods can carry between 5 and 45 ml O_2/100 ml blood. The importance of this increased capacity is clear. If all the O_2 were carried in physical solution instead of in combination with Hb, a man would have to circulate his blood at least 30 times more rapidly than he does.

The sigmoid shape of the O_2-equilibrium curve illustrates important properties of blood as an O_2-carrier. It shows that the amount of O_2 unloaded for a given decrease in P_{O_2} depends upon the value of P_{O_2}. The position of the steepest part of the curve shows where the O_2-content is most sensitive to changes in P_{O_2}. A reduction of P_{O_2} from 100 to 90 mm Hg may for example reduce the O_2-content of blood by less than 0·1 volume per cent while a reduction from 30 to 20 mm Hg may cause 2 volume per cent of O_2 to leave the blood.

The O_2-affinity of blood varies from one species to another. It is often expressed as the P_{50}, that is the P_{O_2} at which the pigment is half saturated. This value lies in the P_{O_2} range where the O_2-content is most sensitive to P_{O_2} changes and coincides with the range of P_{O_2} normally present in the blood.

The O_2-carrying properties of blood are influenced by several factors. Increased H^+-ion concentration will decrease the O_2-affinity of blood of

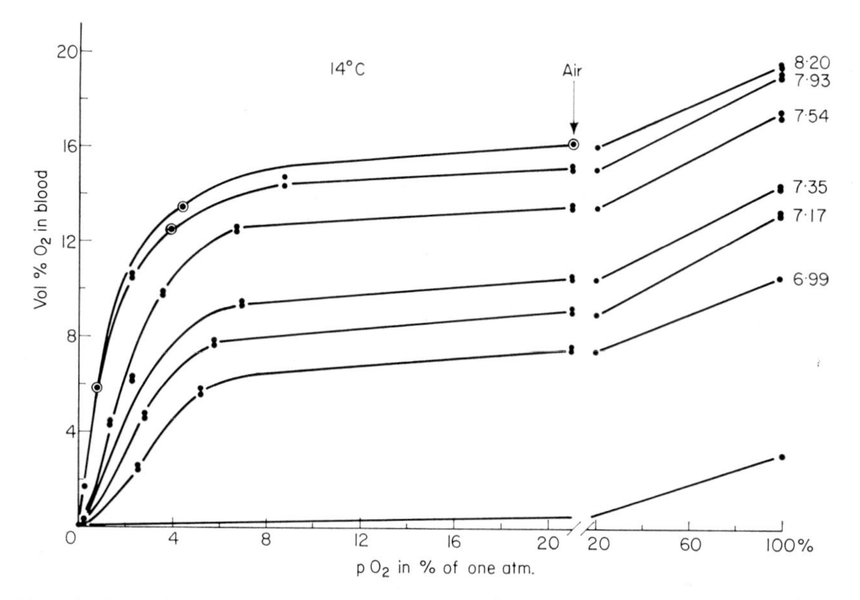

Fig. 1.3. O_2-equilibrium curves of eel blood at 14°C and pH from 6·99 to 8·20. Notice the effect of pH on the O_2-capacity (Root-effect). Bottom curve O_2-equilibrium curve of plasma. Haematocrit of whole blood 40 volume per cent (from Steen, 1963a).

most animals. This was first described by Bohr *et al.* (1904) and is termed the Bohr-effect or Bohr-shift (Fig. 1.2). In some bloods variation in pH will *in addition* change the O_2-capacity. This effect is termed the Root-shift after its discoverer R. V. Root (1931) (Fig. 1.3).

The extent of these shifts is conveniently expressed by a graph of P_{50} versus pH or P_{CO_2}. Bloods with a large Bohr-effect have a very pH-sensitive O_2-affinity, those with a large Root-effect have a very pH-sensitive O_2-capacity. The extent of both the Bohr- and the Root-shifts varies among species and shows adaptations to the habitats and behaviour of the animals (Fig. 1.4). A large Root-shift is always accompanied by a large Bohr-shift, but the reverse is not the case.

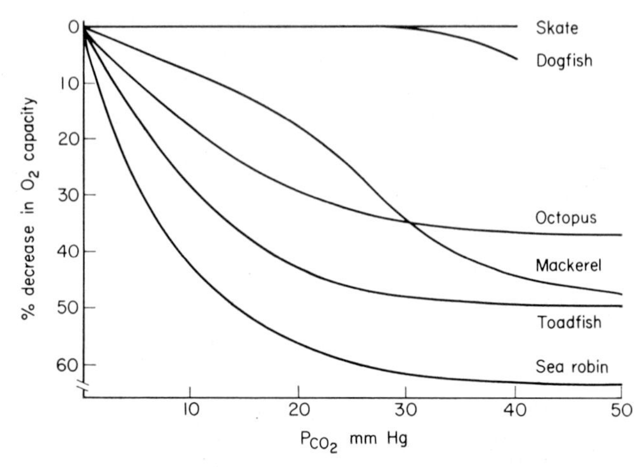

Fig. 1.4. The Root-effect of blood from various fishes (from Lenfant and Johansen, 1966).

When we measure the pH of blood we are, strictly speaking, measuring the pH of plasma. But this pH is not necessarily that inside the red cell, i.e. not that which determines the behaviour of Hb. In some fishes the pH difference across the red cell membrane approaches one unit at high plasma pH (Steen and Turitzin, 1968). Thus graphs of pH versus O_2-capacity of such bloods are markedly different depending on whether they have been obtained on whole or haemolysed blood (Root and Irving, 1943) (Fig. 1.5). This pH difference is most likely a Donnan distribution effected by charges on Hb. The phenomenon has obvious physiological significance and has frequently

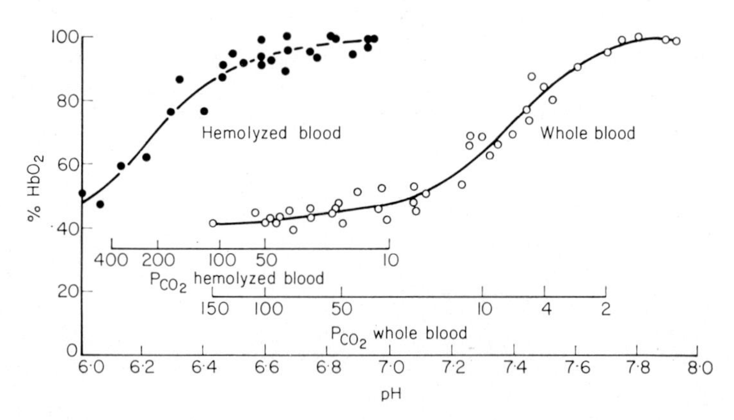

Fig. 1.5. Illustrates the importance of distinguishing the plasma pH from the intracellular pH. The percentage O_2-saturation is used as an intracellular indicator of pH. When the plasma pH is 7·5 (75 % O_2-saturation) the intracellular pH appears to be 6·3. All measurements at 15°C and P_{O_2} = 155 mm Hg. The pH was modified by P_{CO_2} (from Root and Irving, 1943).

been ignored when conclusions regarding the physiological behaviour of blood have been drawn from measurements of properties of Hb or haemolysed blood.

The fact that the O_2-affinity is sensitive to the H^+-ion concentration has considerable importance in normal gas exchange. In the metabolizing tissues, metabolic acids, mostly CO_2 and lactic acid acidify the blood, thus decreasing the O_2-affinity of the blood, i.e. dissociate part of the Hb O_2. This causes increased P_{O_2} in blood plasma and consequently greater ΔP_{O_2} between blood and the tissue. In the respiratory organs the opposite process takes place. Carbon dioxide is given off thereby causing an increased O_2-affinity, a lowered P_{O_2} in the plasma and consequently a higher ΔP_{O_2} between the external medium and the blood. The extent of the Bohr-shift varies from animal to animal and is an important parameter in ecological physiology. This will be extensively treated in later chapters.

The occurrence of a Root-shift seems to parallel that of a swimbladder although it is also found in cephalopods and crustaceans. Its respiratory function will be dealt with in connection with fish respiration (Ch. 3).

Increased temperature generally reduces the O_2-affinity of respiratory pigments. The extent of the temperature effect varies widely among species and correlates well with the temperature fluctuation of the animals.

It is also well known that the O_2-affinity of respiratory pigments is influenced by other aspects of the chemical composition of their environment. For example, the O_2-affinity of haemocyanin is sensitive to the degree of dilution of the pigment, and the O_2-affinity of Hb to the concentration of the commonly occurring ions. However, since the pigment concentration and ion composition of the blood is regulated within narrow limits in living animals, these effects have more biochemical than physiological interest. Quite recently, however, Benesch *et al.* (1968) discovered that 2,3-diphosphoglycerate (2,3-DPG) affects the O_2-affinity of Hb in solution. And the concentration of 2,3-DPG in red blood corpuscles varies with the physiological situation, particularly with acclimatization to altitude (Ch. 6).

Both invertebrates and vertebrates have non-circulating haemoglobin or myoglobin in some tissue cells. The cellular myoglobin has generally a higher O_2-affinity than the circulating pigment and may facilitate O_2-transport from blood to tissue (p. 5).

In addition to its possible transport function, tissue myoglobin can act as a temporary O_2-store, a buffer, to lessen the effect of rapid changes in cellular O_2-uptake.

D. Carbon Dioxide Transport of Blood

The blood also transports CO_2 away from the tissues for disposal at the respiratory organ. Carbon dioxide is transported in the three forms, as physically dissolved CO_2, as bicarbonate and as CO_2 bound directly to Hb as carbaminohaemoglobin.

The reaction of CO_2 with H_2O is described by:

$$CO_2 + H_2O \leftrightarrow H_2CO_3 \leftrightarrow H^+ + HCO_3^-.$$

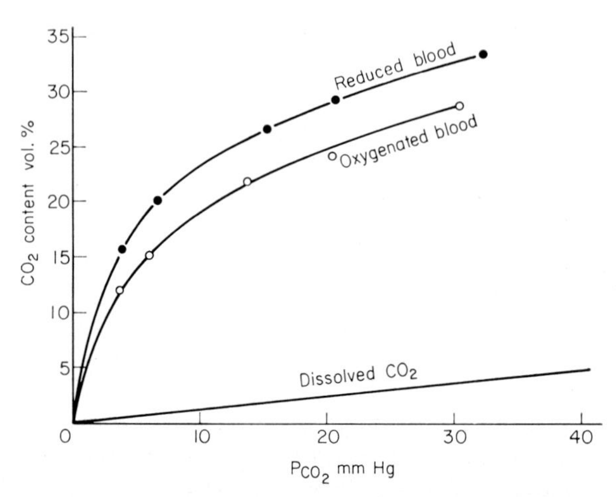

Fig. 1.6. Typical carbon dioxide-equilibrium curves in reduced and oxygenated whole blood from the lung fish *Neoceratodus*. Temperature 18°C. Hematocrit 24 (from Lenfant *et al.*, 1966/67).

At a pH of about 6·2 H_2CO_3 and HCO_3^- will occur in equal concentrations. (The pK of carbonic acid is 6·2.) The lower the H^+-ion concentration, that is the higher the pH, the greater the fraction of CO_2 present as HCO_3^-. Only a small fraction of the total CO_2 is present in physical solution under normal physiological conditions. The amount of CO_2 bound directly to Hb is also relatively small but it may make up one fourth of the A-V difference in CO_2-content. Carbon dioxide is bound to the amino groups of Hb. The amount bound is therefore determined by the pH as well as by the P_{CO_2}.

The largest portion of the total CO_2-content of blood is present as bicarbonate. The actual amount is determined by the ionic composition, particularly the concentration of cations, and by the buffer capacity of the blood. In this connection Hb plays an important role. This has two aspects. One is that Hb itself has a large buffer capacity. This means that it can take up protons, whereby the equilibrium $CO_2 + H_2O \leftrightarrow H_2CO_3 \leftrightarrow H^+ + HCO_3^-$ is moved to the right and more CO_2 is taken up. The other is connected to the *in vivo* situation when uptake of CO_2 occurs parallel to deoxygenation of HbO_2. As HbO_2 is dissociated Hb becomes a weaker acid, and thus a better proton acceptor. Human Hb can take up 0·7 mol H^+ for each mol O_2

given off. This effect of oxygenation of Hb on the CO_2-capacity of the blood is referred to as the Haldane-effect.

The relationship between the amount of CO_2 carried by blood and the P_{CO_2} is expressed in CO_2-equilibrium curves. These are not sigmoid but hyperbolic and usually asymptotic to a line parallel to that expressing the physical solubility of CO_2 (Fig. 1.6).

E. The Rate of Reactions of Respiratory Gases with Blood

The following remarks on the rates of reactions involved in gas transport are included to emphasize the importance of the dynamic state of the ex-

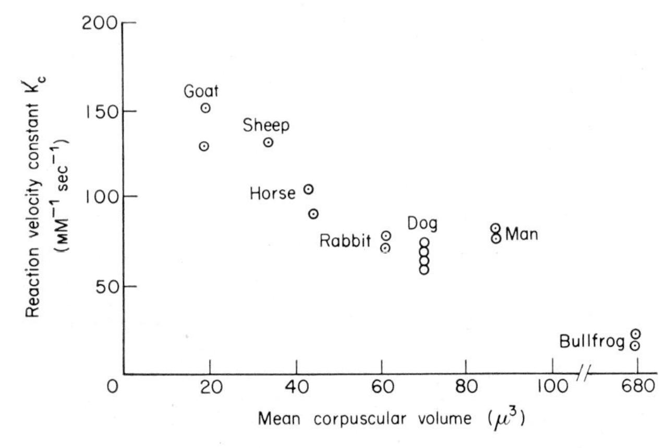

Fig. 1.7. Graph of reaction velocity constant (k_c') for the rate of oxygenation of red cells of varying volume. The small red cells are oxygenated faster than the large ones (from Holland and Forster, 1966).

change process. It can safely be predicted that a full appreciation of respiratory processes depends as much on knowledge about the interplay of the rates of the various processes involved in gas exchange as on the steady state or equilibrium conditions.

Generally speaking the reactions of O_2, CO_2 and CO with blood are too fast to be rate-limiting (Roughton, 1963). In some animals with large red cells, however, these rates are considerably slowed. Holland and Forster (1966) found that bullfrog red cells with a volume of 700 μ^3 take up O_2 some 5 times slower than red cells from goats which have a volume of 20 μ^3 (Fig. 1.7).

The more complex processes, like the Bohr-shift, Root-shift or chloride-shift (the exchange of Cl^- for HCO_3^- across the red cell membrane following CO_2-exchange) may occur at rates which may limit the overall process of gas exchange. The reaction: $CO_2 + H_2O \leftrightarrow H_2CO_3$ is so slow that transport of CO_2 as bicarbonate would be impossible in the absence of an appropriate catalyst. In living cells the reaction is accelerated 1000-fold by the presence of the enzyme carbonic anhydrase. The distribution of carbonic anhydrase in various tissues and among different animals has been reviewed by Maren (1967).

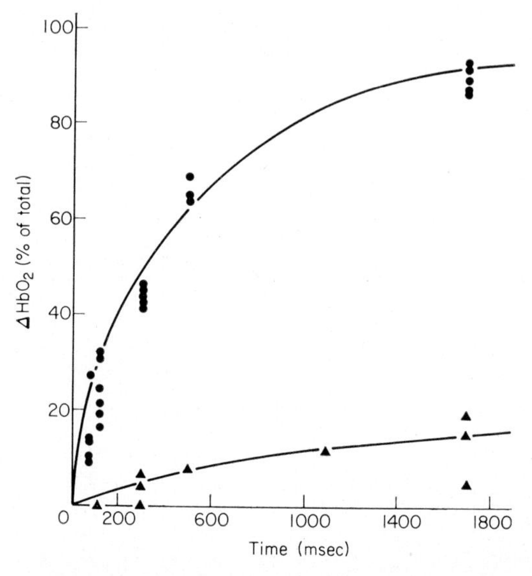

Fig. 1.8. Time course of Bohr off-shift (O_2-flux out of the red cell consequent to acidification) at 37°C created by adding lactic acid to the suspending fluid while keeping P_{CO_2} constant, with (▲) and without (●) 10^{-3}M acetazolamide (from Forster and Steen, 1968).

The enzyme is found inside the erythrocytes and intracellularly in many tissues. Thus the reaction of CO_2 with blood or with a suspension of red cells is very rapid, the half time of the reaction with human red cells in suspension is 2–3 msec. The intracellular location of carbonic anhydrase implies that CO_2 is rapidly converted to carbonic acid which very rapidly dissociates to $H^+ + HCO_3^-$. By this mechanism changes in blood acidity are quickly communicated across the cell membrane. Rapid pH changes are required if the Bohr- or Root-shift shall have functional significance. Craw *et al.* (1963) showed that the half time of the Bohr-shift of human red cells at 37°C is about 0·120 sec. In the presence of acetazolamide, which inhibits the catalytic action of carbonic anhydrase, the half time approaches one minute (Fig. 1.8). Thus, CO_2 is the "carrier of acidity" while carbonic anhydrase is necessary to make its action rapid enough to be of physiological significance. This is well illustrated in an experiment where the pH of a human red cell suspension was changed by lactic acid in the absence of CO_2. This gave a half time of 10 sec for the release of O_2 (Bohr-shift) (Forster and Steen, 1968).

The rate of the Bohr-shift is temperature sensitive (Fig. 1.9). In human red cells the half time is 3 sec at 25°C compared with 0·120 sec at 37°C. The rate of Root-shift was studied by Forster and Steen (1969). When eel red cells are acidified in the presence of CO_2 the half time is 0·050 sec at 30°C. This is appreciably faster than the Bohr-shift of human cells at 37°C. However, at a more normal eel temperature of 15°C the half time was 0·120 sec. When the pH around eel red cells was increased, O_2 was absorbed with a half time of several seconds. This is remarkable compared to a half time of 0·2 sec for

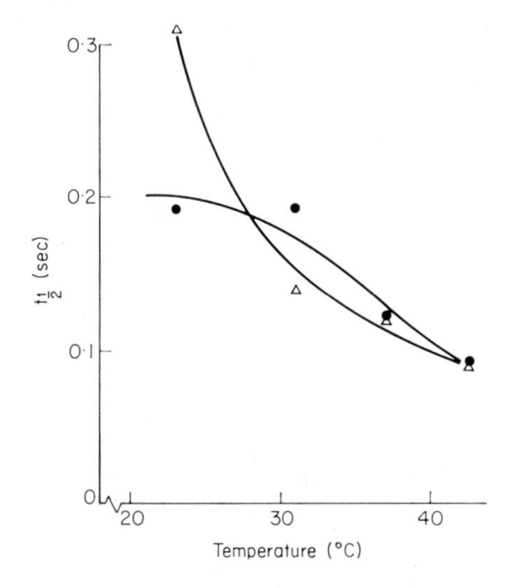

Fig. 1.9. The half time ($t_{\frac{1}{2}}$) for Bohr-shifts created by acidification of red cell suspension at varying temperatures. Circles: initial oxygen saturation 86–88%, triangles: 93–98% (from Forster and Steen, 1968).

the same reaction of human red cells. A detailed discussion of these problems is found in the papers by Forster and Steen (1968 and 1969) and some of its salient consequences are probably manifested in the mechanism of O_2-concentration by the swimbladder.

A further study of the rate of these processes in a variety of animals should be rewarding. Adaptations to environmental and physiological conditions may become apparent. It would also be interesting to compare these rates with the circulation time of blood through the capillaries.

Chapter 2

THE RESPIRATORY ENVIRONMENTS

Animals breath in air or in water. The physical properties of these two media differ in many respects that are essential for their respiratory qualities. These properties have influenced the structure and function of the respiratory organs and possibly also other aspects of the development of animal life in the two media.

A. Air

Air is composed of O_2, CO_2, N_2, traces of other inert gases and water vapour. The relative composition of the gases after removal of H_2O is remarkably constant; 20·95% O_2, 0·03% CO_2, the remaining 79·02% is inert gases of which N_2 makes up about 99·5%. The composition is the same in hot, cold, dry and humid areas. However, since water vapour makes up a varying proportion of the total barometric pressure, the partial pressure of the individual gases may be different. Thus even if cold Norwegian winter air and damp tropical air both contain 20·95% O_2, the partial pressure of O_2 at the same barometric pressure is higher in Norway than in the tropics.

The air on the summit of Mount Everest (8848 m) also contains 20·95% O_2, but since the barometric pressure decreases with the height above sea level, the partial pressure at this elevation is only about 65 mm Hg. Definite adaptations in the respiratory system have been found in animals living at high altitudes (Ch. 6).

B. Water

The solubility of the different gases in water is given in Table 2.1. If water is heated gas solubility decreases and gas will escape, or the gas tension will increase if the water is in a closed system. Cooling will reverse this process. In general gases are less soluble in salt solutions than in pure water. There are salts, however, which in small concentrations increase the solubility of gases in an aqueous solution.

The hydrostatic pressure of water increases by roughly 1 atm per 10 m depth. This increase causes a slight compression of the water. Enns *et al.* (1965) found that the partial pressures of dissolved gases increase by about 14% per 100 atm hydrostatic pressure. Thus water equilibrated with air at the surface was found to have a partial pressure of O_2 of 0·8 atm at 10·000 m depth, if no gas is added or removed. Organisms living at this depth may thus have a very different respiratory situation from those living higher up in the ocean.

While O_2 is only physically dissolved in water, CO_2 occurs, at neutral pH, predominantly as various hydration products. The primary reaction is: $CO_2 + H_2O \leftrightarrow H_2CO_3$. Berger and Libby (1969) have presented evidence

TABLE 2.1

The solubility of O_2 and N_2 in salt and fresh water at temperatures from 0 to 30°C. Solubility expressed as millilitres gas at STPD dissolved in 100 ml fluid at a gas partial pressure of 1 atm (from Fox, 1909 and Rakestraw and Emmel, 1938).

	Oxygen		Nitrogen	
t°C	Fresh water	Salt water (chlorinity 20 wt 0/00)	Fresh water	Salt water (chlorinity 20 wt 0/00)
0	5·05	3·82	1·89	1·79
5	4·43	3·47	1·65	1·59
10	3·92	3·12	1·50	1·44
15	3·54	2·84	1·37	1·30
20	3·22	2·60	1·24	1·20
25	2·96	2·38	1·16	1·11
30	2·73	2·18		

which indicates that this reaction is catalysed at least in some areas of the ocean. Possibly these waters contain an enzyme like carbonic anhydrase or an inorganic catalyst. Carbonic acid dissociates into bicarbonate and hydrogen ions, both of which react further. It is therefore evident that any other ion or substance which will combine with any of these products will influence the equilibrium and thus the apparent solubility of CO_2. When the H^+-ion concentration increases, the equilibrium is shifted toward physically dissolved CO_2. At pH below 4·0 (the pK_a of carbonic acid in water at 25°C is 6·37) almost all the CO_2 is in physical solution. Under these conditions one can therefore obtain values for the physical solubility which can be used to find the fraction of "bound" and "free" CO_2 in other samples for which the CO_2 tension and total CO_2-content are known. In pure water at 15°C the absorption coefficient of CO_2 is about 100 volume per cent at 1 atm P_{CO_2}.

The biological importance of the high solubility and acid character of CO_2 will be discussed later.

C. NATURAL WATERS

Both fresh and salt water are continuously equilibrated with air at the surface. At equilibrium the P_{O_2} and P_{CO_2} will thus be about 155 and 0 mm Hg, respectively. Transport of gases from the surface to deeper layers takes place by diffusion and by convection. Diffusion is a slow process over distances above a few millimetres. It has been calculated that the first O_2-molecule would reach a depth of 250 m in 42 years if diffusion were the only transport process. Convection of water is thus the more important of the two.

Convection takes place to a large extent by currents generated by the wind, but temperature variations also generate convection by influencing the specific gravity of water. Fresh water has its maximum specific gravity at 4°C. The presence of dissolved salts lowers the temperature of maximum density. In brackish water of 18·6‰ salinity the maximum is at 0°C, and in ocean water it is at $-3{\cdot}5°$. These differences have important ecological consequences.

The gas composition of natural waters varies greatly and gas tensions often show sharp stratification even in small lakes and fiords. The P_{O_2} and P_{CO_2} from different localities may, in fact, vary between full saturation and complete desaturation. This is mainly due to the combined results of respiration and photosynthesis. Which of these processes dominates depends on the season, latitude and water transparency. Oxygen contents up to twice that corresponding to equilibration with the atmosphere have been measured. The thickness of the photosynthetic layer depends on the transparency of the water which partly reflects the concentration of organisms and the turbidity due to organic matter. In fresh water ponds light may penetrate only a few metres while in the ocean it can often reach one hundred metres. Below this zone respiration tends to dominate over O_2-production and the O_2-content tends to be low.

Generally it can be said that most rapid rivers are air equilibrated. In the open oceans it appears that the combined effects of photosynthesis, diffusion and convection of water masses are able to maintain an O_2-tension which is ample for animal life. The literature contains numerous exceptions, however, a classical one being the discovery by the Dana Expedition of 1922 of an almost O_2-free layer in the Gulf of Panama.

A very different situation is found in tropical slow-flowing rivers which abound in living organisms and where the turbidity is often high owing to humus. During the 1967 expedition to the Amazon river with the "Alpha Helix" of Scripps Institution of Oceanography, we had a vivid demonstration of this. The water temperature in mid-river was always high, 28–30°C, and the P_{O_2} about 60–80 mm Hg. It is very unlikely that an appreciable part of this river had significantly more favourable respiratory conditions. Thus both the anatomy and respiratory functions of inhabitants of this river must *a priori* be considered as adapted to this environment.

The bottom deposits, both in the sea and in fresh waters, are frequently supplied with oxygen only by diffusion from the water just above, as convection and photosynthesis are practically absent. These deposits always contain large amounts of organic material. Oxygen is consumed mainly by micro-organisms, and generally only the surface layer down to a depth of a few millimetres and sometimes even less, contains free oxygen. Further down anaerobic bacteria reduce oxides and often change the colour of the deposit from a light grey to black. The production of hydrosulphide and methane in these layers is quite common, and in many ponds bubbles of methane and carbon dioxide rise regularly from the bottom and may cause further reduction in the O_2-content and increase the CO_2-content of the water.

D. Biological Consequences of the Difference Between Air and Water as Respiratory Media

The most obvious difference between water and air is that one is a fluid, the other a gas. Thus water has a far higher density and viscosity than air. This implies that it will inevitably require more energy to ventilate an organ by water than by air. The respiratory consequence of this is amplified further by the fact that water equilibrated with air contains only one thirtieth as much O_2 as air and that gases diffuse about 10^6 times faster in air than in water. On this basis it is not surprising that respiratory organs for air and water are principally different. Lungs are ventilated in a tidal fashion, but renewal of gas at the exchange surface depends upon diffusion as well as ventilation. Aquatic gas exchangers are on the other hand usually ventilated in a unidirectional fashion, and the water flow is spread out across the exchange area to provide maximal contact. This basic difference in design implies different functional potentials. Tidal, bidirectional ventilation makes a relatively stable local respiratory environment possible. An example is the alveolar composition which has a lower P_{O_2} and a higher P_{CO_2} than ambient air. This "buffers" the blood gas composition against sudden changes. A similar buffer is more difficult to establish in a unidirectionally ventilated organ where the surface is in direct contact with the ambient medium. Theoretically, a one-way stream of air through air capillaries appears to have certain advantages. It could provide a higher alveolar P_{O_2} and make possible counter-current exchange. The lungs of birds are interesting in this regard. While mammalian lungs are ventilated the bird lung is flushed by air alternately moving from mouth to air sacs and back to the mouth. However, the gas in the respiratory units of bird lungs is not renewed by ventilation, but by diffusion exchange with ventilated air ways. Several benefits can be gained from shielding the respiratory exchange area from direct ventilation. One advantage is that the relatively stable and controllable gas composition of the alveoli makes possible a better regulation of the internal environment. A high alveolar P_{CO_2} gives the organism a basis for employment of the bicarbonate buffer system in acid base regulation. And this system has the unique advantage of having volatile components. This would be almost impossible had the respiratory areas been directly flushed by air. For water breathers the situation would be even worse since CO_2 is about 30 times more soluble in water than O_2, so that an adequately high blood P_{CO_2} is incompatible with a reasonable P_{O_2}.

When an air-breathing animal exposes its blood to the gaseous environment, only dissolved gases and a few other volatile substances are candidates for exchange. Ions can of course not diffuse into the lung gas.

Water breathers are faced with a very different situation. They expose their blood to water and all substances will be exchanged, although to a varying degree. And although CO_2 and O_2 penetrated biological membranes at least 1000 times faster than ions, this accessory exchange is still of significance. This is attested to, for example, by the presence of salt-secreting cells in the epithelium of fish gills. Similar mechanisms for controlling the composition of the internal environment are found among other water breathers.

Still another problem facing water breathers is related to the different heat capacity of air and water. Water breathers expose their blood to an external fluid of the same heat capacity. In contrast the blood of air breathers is exposed to air which has a 3000 times lower heat capacity (on a volume basis). Moreover while 100 volumes of air contain 20 volumes of O_2, 100 volumes of water equilibrated with air contain only one thirtieth of this at 10°C. This means that in order to bring the same amount of O_2 to the respiratory surface, water breathers must expose their internal environment to a heat sink which has a 90 000 times larger capacity than air. This difference leaves no wonder that true homeotherms are never found among water breathers. Homeothermy, which is a major step away from the slavery of the environment, is thus denied to them altogether.

A human bio-engineer might endeavour to make water breathers homeothermic by placing counter-current heat exchangers between the gills and the rest of the body. This principle is in operation in fast-swimming fishes like the tuna in which the blood vessels to and from the swimming musculature pass each other in close contact in counter-current fashion so that metabolic heat is conserved in the muscle. Measurements have shown that these muscles are maintained at about 30°C fairly independently of the temperature of the ambient water (Carey and Teal, 1966).

Gas exchange with air is thus less expensive, and more compatible with a constant composition and temperature of the organism than is aquatic respiration. This makes it easier for air breathers to attain a constant "milieu intern", that is freedom from the environment, than for water breathers. And this independence may have been one reason why the mental capacity has become developed almost exclusively among air breathers. It may, in fact, be inherent in the basic properties of water and air that man should write books on fishes and not vice versa.

Chapter 3

AQUATIC GAS EXCHANGE

Life originated in water. In this chapter we shall describe how the respiratory mechanisms for aquatic gas exchange function. We shall begin with the most primitive animals, where no specialized structures are present for gas exchange, and continue up to the fishes which exhibit specialized and most efficient respiratory organs.

A. Gas Exchange Without Respiratory Organs, Circulatory Systems or Respiratory Pigments

The simplest form of gas exchange is found in Protozoa, Planaria, Rotatoria, many eggs and small embryos. These animals lack organs for gas exchange altogether and gas exchange occurs across the entire cell surface. In many flagellates, however, the surface structures indicate various gas permeabilities for different areas. Experimental evidence on this point is scarce.

August Krogh's (1941) treatment of gas exchange in small organisms is instructive. He imagined an idealized cell of spherical shape, homogeneous surface, constant O_2-uptake throughout the entire cell and P_{O_2} of zero in the centre. Harvey (1928) developed the following formula to describe gas exchange of this cell quantitatively

$$C_O = Ar^2/6D$$

Where C_O is the P_{O_2} at the surface of the cell expressed in atmospheres, A the O_2-uptake in ml/g/min, r the radius of the sphere in centimetres and D the diffusion coefficient. We can take a Paramecium as an example. It has a volume of 0·0006 ml and an O_2-uptake of 1·3 ml/g/hr (Prosser and Brown, 1962, p. 159) i.e. $\frac{1:3}{60}$ ml/g/min and a radius of $5{\cdot}3\times10^{-2}$ cm. Using D of muscle (Krogh, 1941) we get:

$$C_O = \frac{1{\cdot}3\times28\times10^{-4}}{60\times6\times0{\cdot}000014} = \frac{1{\cdot}3\times28}{36\times1{\cdot}1} = \frac{36{\cdot}4}{39{\cdot}6} = 0{\cdot}73 \text{ atm}$$

This external P_{O_2} is incompatible with the conditions of natural waters and indicates that some of the assumptions must be in error. First of all, *Paramecium* is not a sphere, but a bean-shaped body; secondly, and probably most important, it shows considerable protoplasmatic streaming which will enhance gas transport above that achieved by diffusion. Thirdly, the diffusion coefficient may be in error.

This treatment brings out the obvious point that in animals below a certain size sufficient gas exchange will occur by diffusion without any special provisions. But it cannot give accurate quantitative information on the limiting size since no organism behaves like such a model.

Sponges, coelenterates, ctenophores, flatworms, roundworms and echinoderms exhibit some organization of the respiratory function. Sponges show a simple ventilatory system. Water is moved by flagella through numerous pores and irrigates a large internal surface. The diameter of these pores is, however, relatively large, often one millimetre or so. A rough inspection of these animals leaves the impression that the permeability of the outer surfaces may be lower than that of the walls of the pores. This situation is thus principally similar to that found in higher animals where an outer protective surface shields the delicate exchange surfaces inside the animal. Hazelhoff (1939) measured up to 90% O_2-extraction from the water passing through the pores of sponges. However, such values should be used with caution since irrigation is often intermittent. The O_2-extraction during apparently normal breathing was always less than 20%.

B. Gas Exchange with Primitive Respiratory Organs and Circulation, but Without Pigment

Among the echinoderms we find the first unmistakable respiratory organs, and a primitive type of circulatory system, but no respiratory pigment in the internal fluid is present.

Echinoderms have two kinds of respiratory organs. Asteroids (starfishes), Ophiuroids (brittle stars) and Echinoids (sea urchins and their like) all respire by external appendages. These consist of movable tubes protruding out through openings in the outer covering. The soft-bodied sea cucumbers (Holothuroidea) have an internal respiratory tree which is intermittently or rhythmically, always rather slowly, irrigated by muscular movements. The water ventilation serves to transport food and excretory products as well as respiratory gases. Also, it possibly helps to keep the animal turgid (hydrostatic skeleton). This is one of the few examples of a two-way water ventilated respiratory organ. Typically it is found in a primitive animal and serves several functions.

Hazelhoff (1939) investigated the respiration of the sea cucumber *Holothuria tubulosa*. It renews the water of the vascular tree every 1–4 min and the expelled water has an O_2-content of about 50–80% of that in the surrounding water. The O_2-extraction of these animals is thus about the same as for sponges.

The sea urchin can serve as an example of the other types of echinoderms, which perform gas exchange across external appendages. Its anatomy is described by Hyman (1955). It has a thick, calcareous, gas impermeable shell with polar openings for mouth and anus. Around the mouth small finger-like structures protrude; these are often called gills, although their respiratory function is rather doubtful. The main respiratory exchange surfaces consist of numerous podia or tube feet found in five double meridian rows. Each podium penetrates the shell by a two-hole pore (Fig. 3.1). Inside the shell the podium ends in an ampullae. The ampullae from neighbouring podia lie side by side and communicate with a common radial canal. The five radial

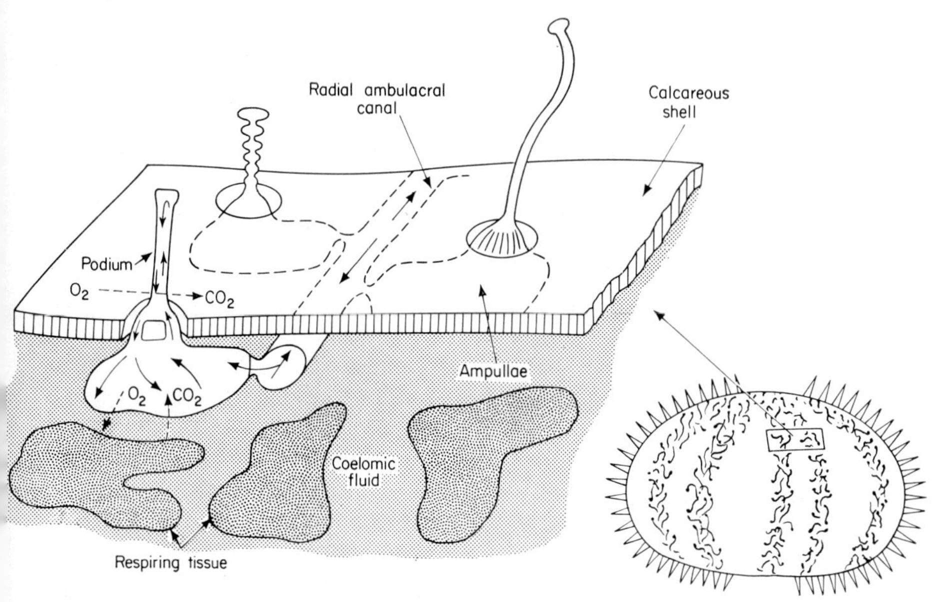

Fig. 3.1. The circulatory system of the sea urchin with section through one podium and its ampullae, connected to the radial canal (from Steen, 1965).

canals connect to a circular canal around the mouth. The fluid inside this system contains no respiratory pigment and is circulated by cilia. Inside a podium fluid moves simultaneously in both directions. The circulatory system of the sea urchin is thus composed of several "microcirculatory" units, each consisting of a podium and an ampullae. There is no capillary system in the tissues. Gas exchange probably occurs directly between the ampullae and the coelomic fluid which fills the internal spaces of the animal not occupied by other organs. No proper ventilation occurs, instead renewal of external medium occurs by movements of the podia. The coelomic surfaces are all ciliated.

The following description (Steen, 1965) amply illustrates the difficulties (and possibly the futility) in defining a primitive respiratory system in quantitative terms.

In a series of experiments the oxygen uptake of the sea urchin (*Strongylocentrotus droebachiensis*) was measured at different temperatures and related to the area of their respiratory surface, which was varied by covering four, six, eight or all ten rows of podia. At 19°C the oxygen uptake was proportional to the uncovered podia. At 6°C, however, four-fifths of the podia could be covered without any measurable effect on the oxygen uptake. This indicates that the respiratory area is a limiting factor for O_2-uptake only at a fairly high O_2-demand. At 6°C the podial surface is apparently four times larger than that required to cover the O_2-demand. One might expect, therefore, that at the high temperature the animal has mobilized all the mechanisms whereby it obtains maximum respiratory capacity. If this is correct a reduction in the external P_{O_2} should reduce the O_2-uptake at 16°C but not at 6°C. This prediction was verified. At 16°C the O_2-uptake was directly proportional to the external P_{O_2} while at 6°C a reduction of external P_{O_2} down to four-fifths of the air value did hardly influence the O_2-uptake at all.

Numerous estimations of the thickness of the podia wall indicated an average of 15 μ. A 70 g sea urchin has about 1000 podia. When fully extended each of them may be 20 mm long and have a diameter of 0·4 mm. This gives an area of $3{\cdot}14 \times 0{\cdot}4 \times 20 \times 1000$ 25,000 mm^2 = 250 cm^2.

To determine the P_{O_2} gradient across the podia membranes one should know the P_{O_2} of the ambulacral fluid as it enters and as it leaves the podia. Such measurements are difficult to carry out. Instead the P_{O_2} of the coelomic fluid can be used. This value will certainly be lower than that of the fluid entering the podia from the ampullae since the coelomic fluid fills the space between the ambulacral system and the respiring tissues (Fig. 3.1). The coelomic fluid had an average P_{O_2} of 0·04 atm.

The ambient P_{O_2} was 0·21 atm. If we assume that the P_{O_2} of ambulacral fluid is 0·04 as it enters the podium and 0·20 when it leaves, the "average" P_{O_2} is 0·12 atm. The "average" P_{O_2} difference between outside and inside will thus be $0{\cdot}21 - 0{\cdot}12 = 0{\cdot}09$. If we use the diffusion constant for muscle tissue of 0·14 ml O_2/min, cm^2 atm/μ (Krogh, 1941, p. 19) in the usual diffusion formula, we get:

$$\text{Vol } O_2 = 0{\cdot}14 \frac{250 \times 0{\cdot}09 \times 60}{15} = 12{\cdot}5 \text{ ml } O_2/\text{h}$$

In comparison the highest O_2-uptake recorded in this investigation was 2 ml/h for a 70 g animal at 19°C.

This discrepancy between calculated and measured respiratory capacity may be reasonably explained by the following factors, although their relative importance is difficult to evaluate.

Firstly, in the calculation we assume all the podia to be fully extended. Direct observations show that this is not the case. It is, however, very hard to find an average "degree of extension". Secondly, we assume the thickness of the podia walls in histological sections to represent the actual diffusion path. The tissue had, however, probably undergone some shrinking during fixation and imbedding. Furthermore, circulation of ambulacral fluid and renewal of external medium may be insufficient, whereby the actual diffusion path becomes larger than the anatomical distance. The lack of separated circulation paths from podia to tissue and back again may also act to render the actual P_{O_2} gradient smaller than assumed in the calculation. There may also be mixing and gas exchange between the fluid streams moving counter-current to each other in the same podium.

The podia of sea urchins serve as locomotory and sensory organs besides their respiratory function. This entails that they are only partially adapted to each of their functions. Such multiple functions of an organ are a primitive condition when compared with more specialized organs like gills. Thus, the wall of the podium is about 15 μ thick as compared to 1–4 μ for the respiratory epithelia of most fishes (Steen and Kruysse, 1964).

C. Gas Exchange with a Respiratory Organ, Organized Ventilation and Respiratory Pigment, but an Open Circulatory System

Many molluscs, crustaceans and annelids fall in this category. In these we find in the majority of cases a specialized respiratory organ with a well-developed ventilatory system, an open circulatory system, often with capillaries in the gills and most important, a respiratory pigment.

Respiration of molluscs. Most aquatic molluscs have gills which serve only for gas exchange. In bivalves, however, the gills also serve to filter and transport food particles, whereas in cephalopods ventilation serves to propel water for locomotion. The circulatory system is more advanced than in echinoderms in that blood is moved by a heart and not by cilia. Energetically this is probably a more efficient mechanism and has the apparent advantage that the smallest vessels can be specialized for exchange. Oxygenated blood is pumped from the heart through discrete vessels to the different tissues. Except for the cephalopods no capillary network is found in the tissues and the blood flows more or less at random back to the heart.

The molluscan gills are built on the same principle that we find in vertebrate gill-breathers. The respiratory surface is increased by foldings. The gills are supplied with blood via an artery and drained by a vein. The external medium is pumped across the gills but the local currents of the exchange surface are controlled by external cilia. There is a counter-current flow of blood and water. That is to say: blood flows inside the flat lamellae in an opposite direction to the flow of water between the lamellae. Such an arrangement enables the arterial P_{O_2} to approach that of inflowing water. The molluscan gills exhibit another typical feature in that they are rather soft and collapsible, although in some, supporting rods of chitin are present in the lamellae. The thickness of the water to blood barrier in individual lamellibranchs varies from 5–100 μ (Atkins, 1936), but a meaningful average is hard to obtain.

The gill surface area has been estimated for representatives of several classes of molluscs (Table 3.1). It varies between 7 and 13 cm^2 per g net weight (Pelseneer, 1935; Yonge, 1947), which is within the range found for gill area of teleosts (Gray, 1954; Steen and Berg, 1966).

TABLE 3.1

Respiratory surface in some adult molluscs (from Pelsener, 1935).

Class	Species	Body weight (g)	Number of lamellae per gill	Total surface (cm^2)	Cm^2/g wet meat
Amphineura	*Chiton pellis serpentis*	2·73	70	23·6	8·66
Gastropoda	*Trochus cinerarius*	0·35	225	2·88	8·64
	Petella vulgata	7·26	272	68·0	9·36
	Buccinum undatum	20·0	256	158·7	7·94
	Purpura lapillus	1·05	106	7·45	7·1
	Helix pomatia (lung)	13·0	—	107·5	8·3
Bivalvia	*Mytilus edulis*	12·0	—	108·0	9·0
	Cardium echinatum	11·8	122	107·5	9·12
Cephalopoda	*Nautilus macromphalus*	135·0	45–52	1257·0	9·3

A respiratory pigment is present in many molluscs, but the heterogeneity of the group is reflected by the fact that while most have haemocyanin, some have haemoglobin and some have both, while others have no pigment at all. The presence of such pigments introduces a most important respiratory variable (Ch. 1) whose properties are subject to genetic adaptations to life in different environments.

Haemocyanin is present in cephalophods, amphineurans, many gastropods, but not in bivalves. It is always present in solution and never carried in cells. Its concentration, and therefore the O_2-capacity of the blood, is probably limited by the accompanying viscosity and colloid osmotic pressure. The molecular weight of molluscan Hcy is 25 000 on the assumption of one Cu atom per molecule. Since, however, one molecule of O_2 is bound between two Cu atoms, molluscan Hcy carries one mol O_2 per 50 000 atomic mass units of pigment. In blood from most molluscs the pigment is found in aggregates of molecular weight between $2{\cdot}5$–$9{\cdot}0 \times 10^6$. Virtually nothing is known about its biosynthesis.

The O_2-dissociation curves of haemocyanin bloods are usually sigmoid, but hyperbolic curves have been found in blood from some species (Ghiretti, 1966b). The respiratory behaviour of Hcy varies a lot. Blood from cephalopods appears to be very sensitive to pH variations (marked Bohr-shift), to temperature and to the ionic composition of the blood. Wolvekamp *et al.* (1942) found, for example, that P_{50} of *Sepia officinalis* blood was 3 mm Hg at pH = 7·97 but 70 mm Hg at pH = 7·24. The O_2-capacity of Hcy blood rarely exceeds 5 volume per cent, the highest values being found among the most active species.

Haemoglobin (Hb) is found among all classes of molluscs except for the cephalopods and monoplacophorans. In amphineurans, gastropods and scaphopods the pigment is mainly present intracellularly in tissues. In some bivalves Hb is found in the haemolymph as well. Circulating Hb is present

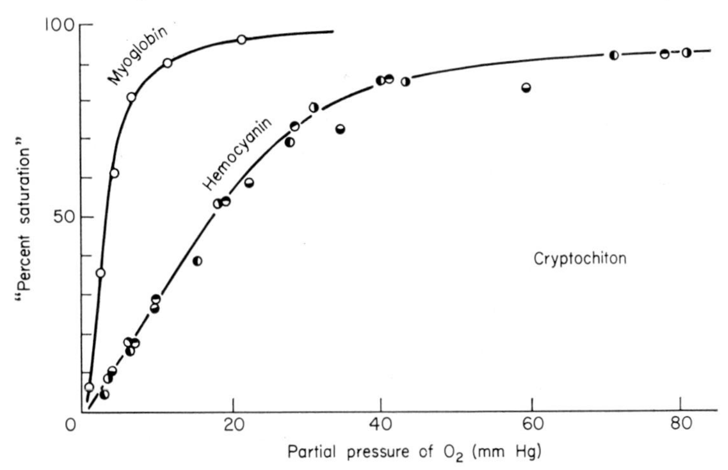

Fig. 3.2. Oxygen equilibrium curves of the myoglobin and haemocyanin of the amphineuran mollusc *Cryptochiton stelleri*. General comments: temperature = 10°C; both pigments from the same specimen (from Manwell, 1958).

Explanation of symbols:

Radular myoglobin

○pH = 7·3; potassium phosphate buffer; final ionic strength = 0·2.

Haemocyanin

Undiluted haemocoelic fluid under constant partial pressure of carbon dioxide (P_{CO}t.)

◐pH = 6·84; P_{CO_2} = 25 mm Hg.

◒pH = 6·42; P_{CO_2} = 75 mm Hg.

3 parts haemocoelic fluid: 1 part potassium phosphate buffer. No carbon dioxide.

◑pH = 7·30. ◓pH = 7·10.

either in solution or, more seldom, in erythrocytes. The latter often has a geometry similar to that of mammalian red cells. Thus Sato (1931) found flat 1 μ thick erythrocytes with 18–21 μ diameter in the mussel (*Area inflata*). In some cases there is Hcy in the haemolymph, but myoglobin in the muscle cells. Manwell (1958) reports a lower P_{50} for myoglobin (3 mm Hg) than for circulating Hcy (20 mm Hg) from the chiton (*Chryptochiton stelleri*). (Fig. 3.2). This suggests a facilitatory function of myoglobin on O_2-transport to the muscles in addition to the storage and buffer function. The fact that the concentration of myoglobin in the radula muscle of clams increases with its volume, further supports such a function (Manwell, 1963).

The molecular weight of molluscan Hb is about 17 000–35 000 when the pigment is present inside erythrocytes. In a few cases, the bivalve *Cardita floridana* is an example (Manwell, 1963), haemoglobin occurs in solution. Interestingly enough it is then present as aggregates with a molecular weight of about 3×10^6.

The Hb content of molluscan blood varies greatly. Fox (1955) showed that in young gastropods (*Planorbis corneus*) a low environmental P_{O_2} stimulates Hb synthesis. In several species the Hb content is greater in mature and larger specimens than in young and small ones of the same species.

The Hb content of molluscan blood is generally low, the highest values rarely exceed 6 mg per cent (human blood contains about 14 mg per cent

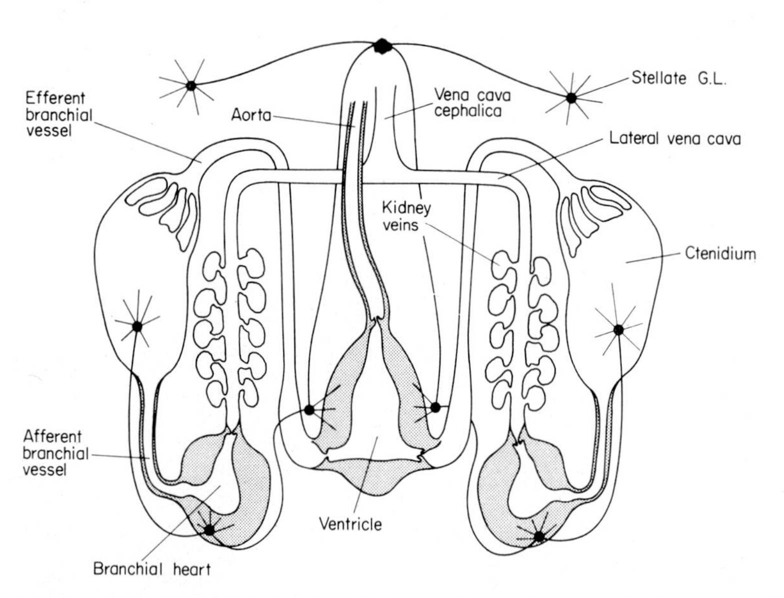

Fig. 3.3. Simplified schematical drawing of the central vascular system in the octopus. The drawing is not anatomically complete (from Johansen and Martin, 1962).

and can combine with 20 volume per cent O_2). However, the content of muscle Hb may often exceed that of human skeletal muscle. Rossi-Fanelli and Antonini (1957) found up to 6 mg per cent Hb in the buccal muscle of *Aplysia deplians*, while human skeletal muscle contains 2·5% myoglobin.

The O_2-affinity of molluscan Hb is difficult to measure since it is very sensitive to the methods of preparation. Values on anything but whole blood have therefore limited interest in the discussion of the *in vivo* respiratory mechanism. According to Read (1966) the P_{50} lies between 2 and 20 mm Hg and is very sensitive to changes in pH and temperature.

The respiratory mechanism of the cephalopod, *Octopus dofleini*, has been thoroughly investigated. Its circulatory system was described by Johansen and Martin (1962). Fig. 3.3 shows a schematic presentation of the central vascular system. From a respiratory point of view the most interesting features are the branchial hearts. They are supplied by deoxygenated blood from the vena cava and supply the gills (ctenidia). From the gills blood enters the ventricle and is delivered to the aorta. This arrangement is analogous to the mammalian system. Pressure recordings from venous vessels further strengthen this similarity: in fact, a high-pressure systemic circulation and a low-pressure branchial circulation exist. This allows the gills to be thin walled and delicate, thus well suited for diffusion, while the systemic circulation can supply blood rapidly to any organ. The same authors also demonstrated good co-ordination between cardiac activity and ventilation. The branchial hearts stop during ventilatory pauses. They also show that during activity the frequency of heart beats increased, and Johansen (1965) showed further that the cardiac output during rest is 10 ml/kg/min while during activity about twice this value. This increase was about parallel to that of the O_2-consumption.

In cephalopods the ventilation principle is intermediate between the bi-directional one of the lung breathers and the unidirectional one of fishes. In cephalopods inspired and expired water flows through separate openings, but some mixing probably occurs inside the mantle cavity. Some authors argue that this obviates an effectively functioning counter-current arrangement. There do not seem to be any direct observations to show the extent of mixing of oxygenated and deoxygenated water. We should be aware that it is the direction of flow at the respiratory surface itself which is of primary importance and this may be effectively directed by the cilia. It seems more realistic to assume that the effect of counter-current flow may be diminished, rather than abolished, by partial mixing of the water.

Van Dam (1954) observed large gradients in the expiratory water from scallops. Thus samples taken simultaneously but at different positions in the effluent water gave O_2-extractions ranging from 33 to 72%. This shows both that the gills may be very efficient and that water is unevenly distributed over the respiratory surface. In this particular species this is most likely a consequence of the dual function of ventilation. However, a similar situation certainly occurs to some extent in all respiratory organs. In octopus arterial P_{O_2} was consistently 10–20 mm Hg below that in effluent water (Johansen and Lenfant, 1966). However, in one case out of twenty-five the arterial P_{O_2} was higher than the effluent P_{O_2}. If the measurement is correct this strongly indicates a functional counter-current exchange. In judging the performance of a respiratory mechanism it is frequently as meaningful to consider the extremes as the average values, since this demonstrates the limit of performance.

As mentioned earlier, water is not evenly distributed across the exchange surface. Johansen and Lenfant (1966) also have evidence for a variable degree of blood shunting in the gills. So when the average P_{O_2} of effluent water approaches that of arterial blood, as closely as indicated above, it is reasonable to believe that the P_{O_2} of that part of the water that has been in the closest contact with blood very nearly approaches that of arterial blood.

The respiratory properties of octopus blood and the general pattern of gas exchange have been investigated by Lenfant and Johansen (1965) and Johansen and Lenfant (1966). In a 9 kg specimen the blood had an O_2-capacity of 3·8 volume per cent. Arterial saturation was 100%, venous saturation 21% at 90 and 12 mm Hg partial pressures, respectively. At these con-

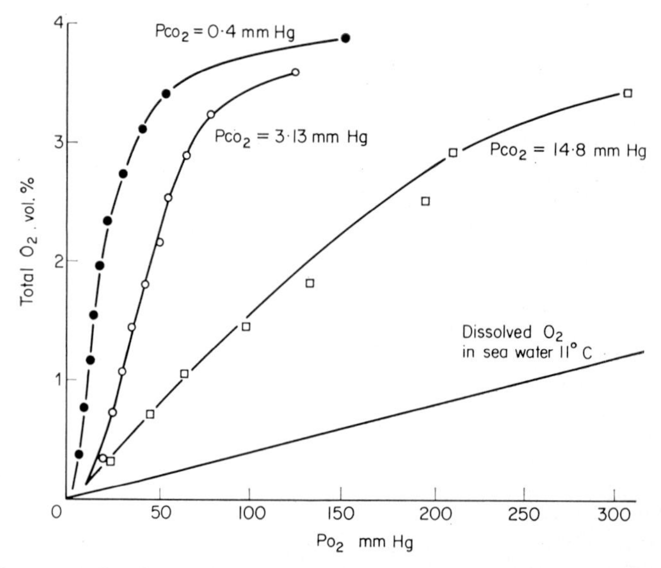

Fig. 3.4. Representative O_2-equilibrium curves at different values of P_{CO_2}. Temperature of the blood was 11°C. Total O_2-content is plotted on the *ordinate*. No plateau is reached because the dissolved oxygen will increase in proportion to P_{O_2} and become parallel to curve at bottom indicating the amount of oxygen dissolved in sea water (from Lenfant and Johansen, 1965).

ditions the heart rate was 10 beats per minute with a stroke volume of 18·1 ml. Percentage O_2-extraction was not measured in this case, but from similar experiments a value of about 25% appears likely. The difference in O_2-content between arterial and venous blood was 3 volume per cent, which is not much below that found in fishes and land animals. The water to blood barrier of the respiratory surface is between 1–5 μ (personal observations), but the area of the respiratory surface is not known.

This octopus lives along the Pacific coast of North America where the water is cool and well aerated. Typically, its blood has P_{50} of about 40 mm Hg at arterial P_{CO_2} which is from 3 to 4 mm. At venous P_{CO_2} (5–6 mm Hg) P_{50} is about 55 mm Hg (Fig. 3.4). This large Bohr-effect thus has significant influence on the gas-carrying capacity; it increases the affinity of blood at the receiver end (the gills) and reduces it at the delivery end (tissues). This entails, in both instances, an increased diffusion gradient for O_2.

Octopus maintains a rather constant blood pH. Thus Johansen and Lenfant (1966) found arterio-venous pH differences between 0·01 and 0·12 with an average of 0·05 pH units.

In comparison Redmond (1962) found blood of four chitons to have a rather pH-insensitive O_2-affinity and a P_{50} of about 20 mm Hg at 25°C. However, two of these species were collected from Jamaica, where they live in 28°C water, the other two from Seattle at a water temperature of 10°C. At these temperatures the P_{50} was 30 and 8 mm, respectively. This is a strange situation since the warm water environment most likely is the O_2-poorer of the two. And it is generally found that species in such habitats have a high O_2-affinity compared to their relatives in more O_2-abundant water. In these chitons the opposite is the case.

The rather stable concentration of Hcy in, for example, cephalopods attests to its respiratory importance. In four species of abalones (*Haliotes*), Pilson (1965) found a 900-fold variation of Hcy concentration within the abalone *Haliotes corrugata* (average 0·15 g/100 ml). This impressive variability indicates that the pigment does not have a crucial respiratory function in these animals.

Respiration in crustaceans. Among crustaceans we find animals with highly specialized respiratory organs. They are distinguished from those of cephalopods, however, in important respects. The structure of crustacean gills is summarized by Wolvekamp and Waterman (1960).

In crustaceans the respiratory surfaces are not ciliated, but covered by a layer of chitin. This gives them considerable mechanical strength and also other properties which we shall discuss later. There is an open circulatory system and the blood volume constitutes 10-50% of the animal's volume. The heart pumps blood to the gills where it perfuses the gas exchange capillaries and returns to the tissues. There is no capillary system in the tissues and blood appears to return to the heart region more or less at random.

The gills are of two types, lamellar consisting of a folded surface, or vesicular consisting of separate tubes. The area of the respiratory surface has been measured in a few crabs (Gray, 1957; Hughes *et al.*, 1969) and is from 300 to 1100 mm^2 per g body weight. This compares well with values for the area of fish gills relative to the weight of the fishes.

Most crustaceans have a respiratory pigment, some have haemoglobin (formerly called erythrocruorin) but most have haemocyanin. Both pigments occur in solution in the haemolymph but intracellular Hb is also known in a few cases (Goodwin, 1960).

Crustacean haemoglobins have molecular weights usually in the range about 400 000; apparently polymers of a basic unit consisting of one protoheme with one Fe^{2+} and a molecular weight of about 17 000. The P_{50} is usually below 3 mm Hg, and the Bohr-effect too slight to be of respiratory significance at physiological A–V pH of P_{CO_2} changes (Fox, 1945). Carbammate formation does not appear to be important in CO_2-transport.

In many crustaceans belonging to the *Cladocera* a most interesting respiratory adaptation has been demonstrated in a series of papers by Fox and co-workers (see Fox, 1955). When living under hypoxic conditions several species of *Daphnia* show increased Hb-concentration. This has been shown to have considerable survival value (increased life span and raised egg production) and the effect is reversed under normoxic conditions.

This adaptation has two interesting aspects besides its function in these organisms. One is that similar causal relationships never seem to occur between Hcy-synthesis and P_{O_2}. The other is that this type of response to hypoxia is typical for Hb-carrying animals all the way up to man. The responsiveness of the Hb-synthesis may have been an important selective factor in evolution. Nothing appears to be known about the flux of "P_{O_2} information" from the environment to the Hb-synthesis, neither is the inability of the Hcy-synthesis to respond to P_{O_2} understood. These basic problems require a rather imaginative approach.

Crustacean haemocyanins usually have molecular weights from 700 000 to 900 000 (Goodwin, 1960). Lobster (*Homarus americanus*) Hcy has a molecular weight of 74 000 per O_2-molecule and occurs in polymers of molecular weight 825 000. It normally occurs in concentrations up to 6 mg per cent. Crystalline Hcy of *Palinarus* has an O_2-capacity of 25 ml O_2/100 g pigment (man: 134 ml O_2/100 g Hb). The O_2-capacity of crustacean blood thus rarely exceeds 1·5 volume per cent.

We may recall that the O_2-capacity of Hcy containing blood of molluscs often approaches 5%. It is difficult to find a rational explanation why crustacean blood consistently shows a lower O_2-capacity. It may be connected to the roughly 10-fold lower molecular weight of crustacean Hcy compared to molluscan. This most likely entails a higher colloid osmotic pressure and viscosity of crustacean than of molluscan blood at equal O_2-capacity. This explanation again raises the question: why do crustaceans not make Hcy with a higher molecular weight?

Prior to the investigation of Redmond (1955) on the respiratory function of crustacean Hcy, the low O_2-capacity of their blood had caused doubt as to the respiratory function of this pigment. Redmond's work definitely established the importance of Hcy in at least some species. He studied the spiny lobster, *Panulirus interruptus*, the sheep crab, *Loxorhynchus gradis*, the lobster, *Homarus americanus* and a small crab, *Pachygrapsus crassipes*. They all had blood with P_{50} from 6 to 8 mm Hg at their normal temperature (15°C) and blood pH (about 7·50). The P_{50} was in all cases very sensitive to variations in pH and temperature (Fig. 3.5).

This is a very high O_2-affinity for animals living in well-aerated water, normal P_{50} values for molluscs and fishes from similar habitats are 15–

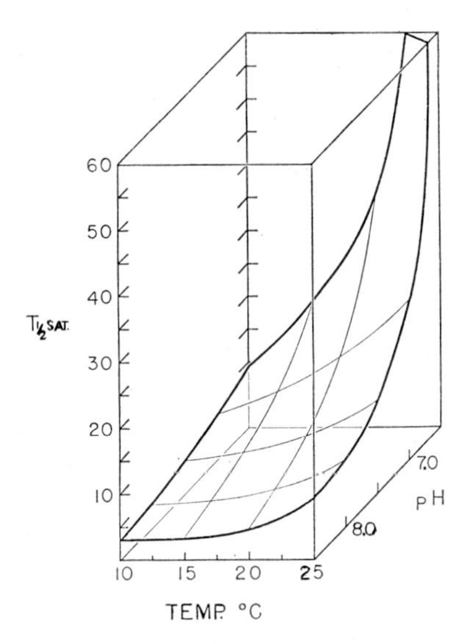

Fig. 3.5. The effect of temperature and pH on the half-saturation pressure of the haemocyanin of *Panulirus interruptus* (from Redmond, 1955).

30 mm Hg. The low P_{50} of crustacean blood makes it technically difficult to investigate possible relationships between the normal environmental P_{O_2} of different species and the O_2-affinity of their blood.

One of the most interesting of Redmond's observations was the conspicuously low arterial O_2-saturation. Post branchial or arterial blood only occasionally had a P_{O_2} above 12 mm Hg which corresponds to about 60% saturation.

The arterial O_2-content was usually about 0·4–0·8, the venous 0·15–0·40 volume per cent. However, since the arterial P_{O_2} is so low, the amount of dissolved O_2 is low, and 80–95% of the O_2 delivered to the tissue is carried by Hcy.

Johansen *et al.* (1970) investigated gas exchange across the gills of three species of large crabs. Their results show that gas exchange in these crustaceans is very different from the picture described by Redmond (1955) for the lobster. The O_2-capacity was 3–4 volume per cent, the P_{50} about 20 mm Hg and arterial saturation almost complete with a P_{O_2} of about 100 mm Hg. Johansen *et al.* used internal catheters. Redmond took samples and analyzed them. This technique may be risky because of the low O_2-capacity of lobster blood. The different methods may be one reason why their results are so different. It would be interesting to study gas exhange in the lobster by internal catheters and thus decide whether gas exchange across its gills really is as remarkable as it appears now.

The properties of the gas exchange mechanism described by Johansen *et al.* for crabs is much like that described earlier for molluscs, and in fact also for fishes.

The gills of crustaceans are paired organs situated laterally under the carapace. Water enters through pores around the edge of the carapace and is expelled through an anterior opening. The direction of flow is frequently reversed for short periods of time. Ventilation is created by a paddle-like structure, the scaphognathite, situated anteriorly in the branchial chamber. Hughes *et al.* (1969) investigated the pattern of water flow in the branchial chamber of the crab *Carcinus maenas* by adding a dye to the inhaled water and observing its distribution through a plastic window in the shell. These observations were supplemented by P_{O_2}-measurements of water at different localities. They confirmed earlier observations that water and blood flow counter-current to each other. The O_2-extraction varied from 10 to 25% from one area of the gill to another.

This O_2-extraction is much lower than demonstrated for a number of other crustaceans, by Lindroth (1938) and Hazelhoff (1939). Differences in the experimental situation may account for the deviations. The only measurements that were performed without disturbing the animal in any way were those of Hazelhoff on the crab *Caloppa granulata*. The O_2-extraction in these experiments ranged from 60 to 90%. The low values reported by Hughes *et al.*, 1969, may thus be caused by the irritation or excitement which was induced by replacing part of the carapace with a plastic window.

D. Gas Exchange with Specialized Respiratory System, Organized Ventilation, Closed Circulatory System and a Respiratory Pigment in Blood Cells

1. THE STRUCTURE OF THE GILL APPARATUS

Fish gills are distinguished by a higher degree of regularity and specialization than the respiratory organs of lower water breathers. Teleost and elasmobranch gills have the same basic geometry and both operate strictly as counter-current exchanges (Fig. 3.6 A and B). Typically in teleosts the 4 gill arches on each side of the pharynx each support two rows of filaments. The

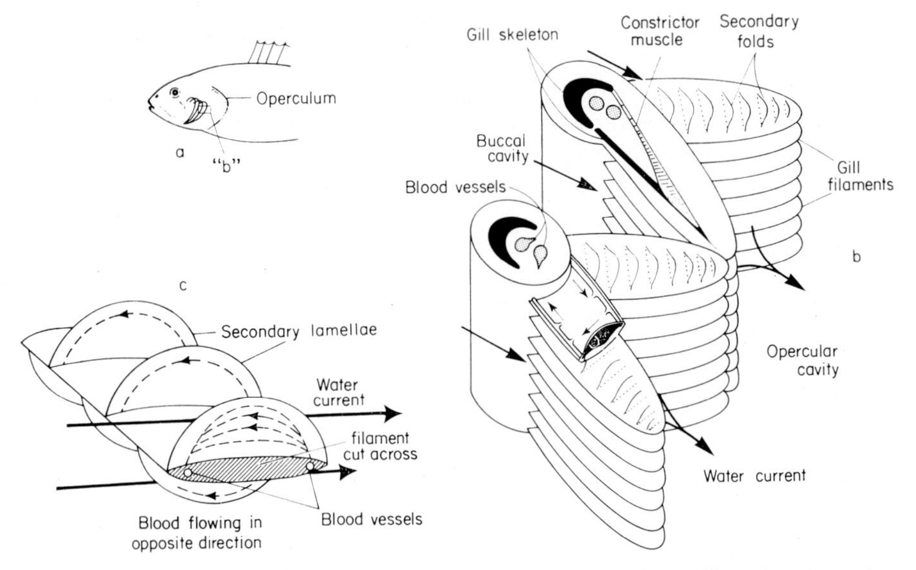

Fig. 3.6A. a (above). Diagram to show the position of the four gill arches beneath the operculum on the left side of a teleost fish.

b (right). Parts of two of these gill arches are shown with the filaments of adjacent rows touching at their tips. The blood vessels which carry the blood before and after its passage over the gills are shown.

c (below). Part of a single filament with three lamellae on each side. The flow of blood is in the opposite direction to the water (from Hughes, 1961).

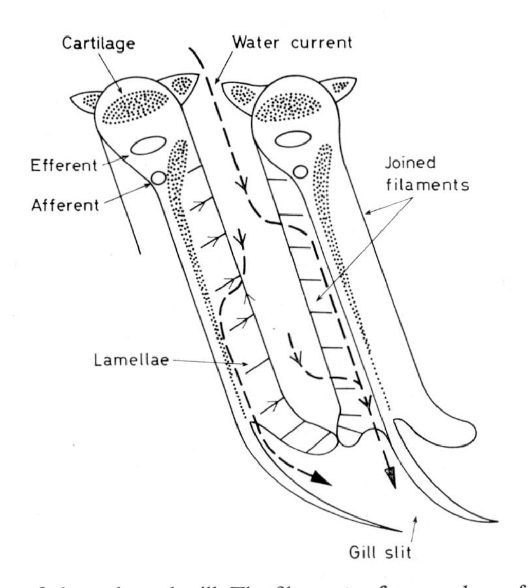

Fig. 3.6B. Diagram of elasmobranch gill. The filaments of one arch are fused.

surface of the filaments is folded into thin-walled, parallel, evenly spaced lamellae. The long axis of the lamellae is perpendicular to that of the filament. The outer edge of lamellae from adjacent filaments touch each other. The lamellar surface comprises the respiratory surface. The position of the filaments, and thereby of the lamellae relative to the water current, is regulated by muscles controlling the angle between the two rows of filaments on the same arch. The tips of the filament on one arch touch those from the adjacent arch. Therefore all the water passing the gill apparatus may ideally flow between the lamellae. Thus the gill structure provides a maximum common surface between the water and the blood circulating the lacunar interior of the lamellae.

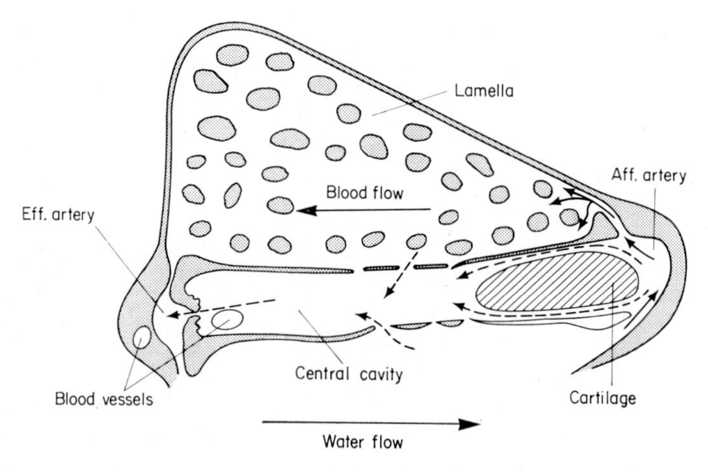

Fig. 3.7. Drawing of a gill filament seen in cross-section with the vascular channels shown with arrows. Solid arrows indicate the respiratory path, dotted arrows the non-respiratory path through the central compartment. The drawing is simplified to emphasize the vascular alternatives (from Steen and Kruysse, 1964).

The branchial artery, carrying mixed venous blood from the heart, gives one branch to each of the gill arches. Each of the gill arch arteries gives one branch to each filament. The filament artery follows the posterior edge and gives one branch to each lamellae. In the lamellae the blood is spread out in a thin film between the two sides of the lamellae, which is held together by pilaster cells. This circulatory pathway is the respiratory route. The respiratory filament arteries communicate not only via the lamellae, but also through open lacunae of lymphatic nature in the centre of the filament (Fig. 3.7). Thus, there exists in the gill filament alternative perfusion paths; a respiratory and a non-respiratory. This was first described by Riess (1881) and was rediscovered by Steen and Kruysse (1964). Possible functional significance of this arrangement will be described later in this chapter.

The fine structure of the lamellae of the pollack (*Gadus pollachius*) has been studied electron microscopically by Hughes and Grimstone (1965) (Fig. 3.8). They also review the pertinent literature. The external surface of the lamellae consists of flat epithelial cells 0·4–1·5 μ thick. These contain mitochondria and an abundance of vesicles, tubules, granules and vacuoles. Beneath the epithelium is a 0·3 μ thick basement membrane. The two lamellar faces are held together by the pillar cells (sometimes called pilaster cells). There are also bands of connective tissue running transversely through these cells. Commonly the nuclei of the epithelial cells lie adjacent to the pillar cells. Thus the nuclei do not lie in the shortest diffusion path. The blood space is delimited by the pillar cells and by their 0·1–0·3 μ thick flanges, which spread out and connect with each other.

The presence of pillar cells is the most unique histological feature of the lamellae and may be considered to be analogous with the endothelial cells of capillaries.

The elaborate structure of the epithelial layer indicates that it may serve other functions besides constituting a diffusion barrier between blood and water. Most likely they have some salt secretory function. Active gas transport is most unlikely. In other species the epithelium is much thicker and at least in these it appears likely that non-respiratory processes takes place. The lamellae of some fishes have cells of distinctive secretory type, which is thought to be specialized for salt transport (Parry, 1966).

2. FUNCTIONAL ANALYSIS OF THE GILL APPARATUS

The gills are sufficiently regularly built to allow quantitative estimations of the functional consequences of the structural parameters. The number of filaments and the frequency of lamellae along them can be measured. Similarly the thickness of individual lamellae and of the space between them can be obtained. The area of individual lamellae can also be obtained. By combination of these measurements the total lamellar surface is obtained. Such studies have been carried out on a number of fishes (Gray, 1954; Hughes, 1966). A critical consideration of various procedures for obtaining gill areas was given by Muir and Hughes (1969). By histological preparation of gills one can also get an idea of the thickness of the blood–water partition (Steen and Berg, 1966).

a. *Ventilation*

The mechanism of ventilation is principally the same in all fishes investigated (Hughes, 1960). It is based on the interplay between a buccal pressure pump and an opercular suction pump. Attempts to classify fishes according to the mode of ventilation has not contributed new concepts. In most pelagic, free-swimming fishes both pumps are reduced, but in some, like the mackerel,

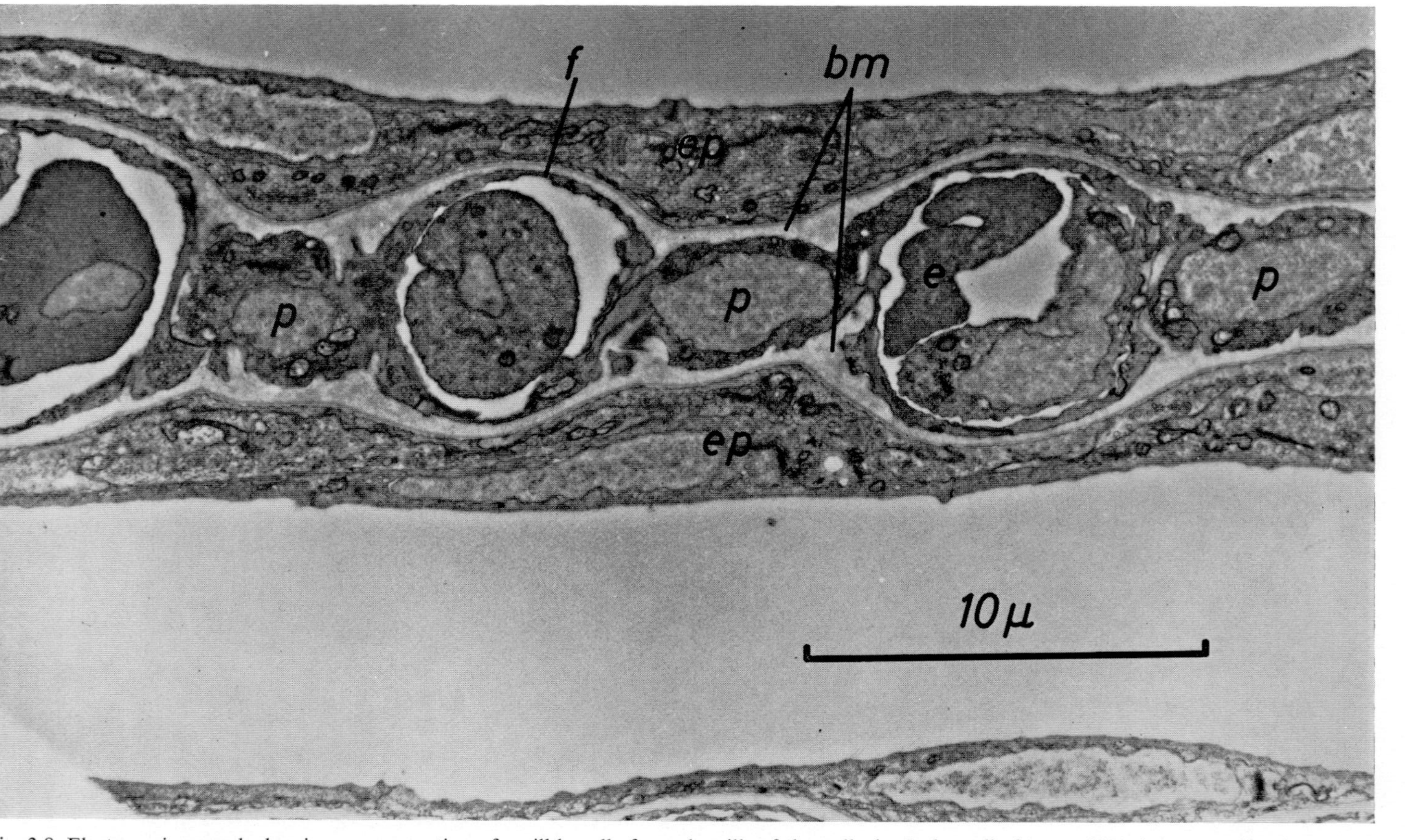

Fig. 3.8. Electronmicrograph showing a cross-section of a gill lamella from the gills of the pollack, *Gadus pollachius*, at 4000 times magnification. Three pillar cells, P, with flanges, f, meeting and joining delimit the blood spaces. Note also basement membranes, bm, and epithelia, ep (from Hughes and Grimstone, 1965).

they are evidently absent altogether. This fits well with the general view that gill ventilation in these fishes is accomplished by using the water current created by their movement. At the other extreme are typical bottom-dwelling fishes which have well-developed pumps. In these the opercular suction pump is particularly well developed and the buccal pressure is higher than the opercular for 90–95% of the time which means that they maintain an almost continuous stream of water, although of rhythmically varying velocity, through the gill sieve. The same situation is probably achieved in swimming fishes simply by keeping the mouth open. The functional significance of a continuous water current through the gill apparatus is obvious.

Hughes (1966) calculated flow through gill models of different dimensions. The usual Poiseuilles's equation for flow through a circular tube was adopted for the case of a rectangular tube, thus making it applicable for flow through the gill sieve model (Fig. 3.9).

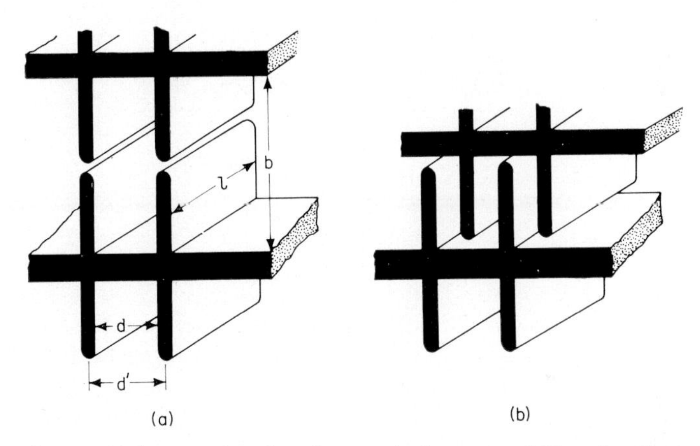

Fig. 3.9. Diagram of the secondary lamellae attached to two neighbouring filaments of the same gill arch to show the dimensions used in the calculations of gill resistance. In (a) the filaments are shown wide apart and with the secondary lamellae opposite one another; in (b) the minimum possible size of pores is shown to result from the close interdigitation of the filaments. It is improbable that such close interdigitation is ever found in life (from Hughes, 1966).

The equation is:

$$q = \frac{p_1 - p_2}{n} \times \frac{5d^3 b}{24l} \qquad (1)$$

where p_1 and p_2 are the pressures (dyne/cm^2) on either side of the sieve (i.e. in the buccal and opercular cavities), n is the viscosity of the fluid in poises, d, b and l are the dimensions in cm of the sieve shown in Fig. 3.9.

This gives the flow q through a single pore. The total flow is obtained by multiplication with the total number of pores (N).

In applying this equation to flow through the gills of the tench, which had been the subject of *in vivo* flow measurements, certain simplifying assumptions were made: (1) All pores of the sieve are equally accessible to water flow throughout the respiratory cycle, i.e. completely even distribution of water. (2) The pressure gradient across the gills remained constant and equal to the mean of the differential pressure. (3) The shape of the pores was assumed to be as in Fig. 3.9.

An example of the use of this equation is given below, where measured values from the tench (Hughes, 1966) have been used.

$$p_1 - p_2 = 0{\cdot}5 \text{ cm } H_2O = 981/2\text{ , say } 500 \text{ dynes/cm}^2$$

$$n = 0{\cdot}01 \text{ poises } (= \text{g/cm} \times \text{sec})$$
$$d = 0{\cdot}025 \text{ mm} = 2{\cdot}5 \times 10^{-3} \text{ cm}$$
$$b = 2 \times 10^{-2} \text{ cm}$$
$$l = 8{\cdot}6 \times 10^{-2} \text{ cm}$$

$$q = \frac{500}{10^{-2}} \quad \frac{5(2{\cdot}5 \times 10^{-3})^3 \times 2 \times 10^{-2}}{24 \times 8{\cdot}6 \times 10^{-2}}$$

$$= \frac{78{\cdot}125 \times 10^{-4}}{206{\cdot}4}$$

$$= 0{\cdot}379 \times 10^{-4} \text{ ml/pore/sec}$$

Number of pores in gill sieve, $N = 26{\cdot}8 \times 10^4$. Therefore the total flow of water through gills, $Nq = 10{\cdot}13$ ml/sec. The normal volume of water pumped through gills of a tench of this size (150 g) is 1–2 ml/sec. A similar calculation for a 50 g *Callionymus* gave values of about 20 ml/sec, whereas the usual volume pumped is about 0·5 ml/sec.

The calculation thus gave higher values than those measured. A review of possible explanations is instructive. The most obvious point where the model differs from reality is the assumption that all pores are equally available to water and of the shape shown in Fig. 3.9. In all fishes the sides of adjacent filaments are joined for a certain length from the base by a septum and water must flow, first past the lamellae and thereafter through channels along this septum. This obviously adds resistance to the water flow. The actual portion of pores of this kind varies among fishes and is not known for the tench. In some elasmobranchs (Fig. 3.6b) only the distal one-tenth or so of the filaments are free. In others, the mackerel is an example, the filaments are free for most of their length.

A further point is that the main flow of water is not necessarily at right angles to the sieve because of the orientation of the filaments in the respiratory chambers. This will add to the resistance as will the presence of gill rakers between the individual arches. Another important feature of the sieve to be considered is the way in which it may modify its position relative to the water current during an individual respiratory cycle. Some of these modifications may be produced passively by the pressure gradient and the flow of water across the sieve, while other changes are produced actively by the branchial arch musculature and also by the abductor and abductor muscles of the individual filaments. The precise nature of the movements of the gills and of the individual filaments during the respiratory cycle is complex and difficult to investigate under normal conditions. Observations on transparent species of *Gobius* have shown that the tips of the filaments tend to be in contact despite the abduction of the operculum, and that there is also a rhythmic adduction of the filaments (Hughes, 1961). Similar conclusions were reached for other teleosts by Pasztor and Kleerekoper (1962) who regarded the first type of activity as a mechanism ensuring even ventilation of the whole sieve and the second as a regulatory mechanism during periods of increased ventilation.

Consideration of the modified Poiseuille equation suggests that the adjustments which might be most important are those which affect the dimensions of the pores through which the water passes. In this connection it must be remembered that the presence of a film of mucus is bound to reduce the effective size of these pores. In the calculation given above it has been assumed that the channels are as shown in Fig. 3.9a but in life the secondary lamellae alternate with one another and may even interdigitate as in Fig. 3.9b. In the extreme case figured, the channels will be reduced in height, *b*, by half and by more than half in thickness, *d*. Calculations based upon such a configuration of the gill sieve gives figures for the flow through the gills of the tench of about 0·07–0·08 ml/sec. This volume is far less than what has been measured (1–2 ml/sec). It is likely, however, that adjacent filaments never interdigitate as completely as has been assumed in this modified calculation. These conditions for which the flow has been calculated therefore represent two extremes. Hence the minute volume of a 150 g tench would be expected to fall within the range 5·0–600 ml which is in fair agreement with the direct measurements (60–120 ml). If the gill resistance changes during the respiratory cycle by variation in the degree of interdigitation of the secondary lamellae it would provide a very effective means of control. This might be accomplished by alterations in the curvature of the branchial arches.

There are, then, many ways in which the gill resistance can vary during the respiratory cycle. The number and dimensions of the channels through which water passes are such that the measured differential pressures are sufficient to maintain the required flow of water across the sieve in such a way that the water is brought into close contact with the respiratory epithelium.

Table 3.2 shows gill measurements on a series of the same species varying in weight from 0·3 to 850 g. The gill area has increased about 1000 times while the calculated total flow with the same pressure has increased by a factor of 350. This, of course, means that the increase in area was accompanied by an increase in resistance per pore. The increased area resulted primarily from an increase in the number of lamellae and of their length (about 50× and 10× respectively). The resultant influence upon gill resistance has been estimated by Hughes (1966).

Using the symbols of Fig. 3.9 the total area is expressed by:

$$A = (2L/d')bl$$
$$= L(2bl/d'). \quad (2)$$

and the total flow by:

$$Q = \frac{L}{d'}\frac{p_1 - p_2}{n}\frac{5d^3b}{241}$$
$$= LK\frac{d^3b}{d'l} \quad (3)$$

where L is the total filament length and L/d' the total number of pores, N.

The only parameter of the gill sieve which is different in the two equations is that of d' and d, but these two are clearly related to one another so that equation (3) may be approximated by:

$$Q = LK(\mathrm{d}^2b/l) \text{ where } K = \frac{5(p_1 - p_2)}{24n}\frac{d}{d'}$$

It is apparent that when one of these parameters is altered the effects upon the flow and upon the area are not necessarily in the same direction. We may summarize the effect of the main parameters on the area, A, and on the total flow, Q, as follows:

(a) increased L — increased A and Q;
(b) increased b — increased A and Q;
(c) increased l — increased A, decreased Q;
(d) increased d — decreased A, increased Q.
(or decreased I/d')

Evidently any increase in the gill area can best be achieved by having longer filaments, each having high secondary lamellae, i.e. increase of L and b.

Table 3.2 shows that this is the way the gill area increases during growth of at least one species. In addition there is an increase in l but this is partly compensated for by a larger distance between adjacent lamellae.

b. Gas Exchange

(1) *General.* The parameters which determine the rate of diffusion across the gill surface (area, gradient, distance and diffusion constant) are to some

TABLE 3.2

Gill dimensions in the fish *Micropterus dolomieu* and water flow through the gill sieve. For details see text (dimensions from Price, 1931, calculations from Hughes, 1966).

Weight	Total filament length (L) (mm)	(d' mm)	Total no. of pores (N) × 10^4	Pore dimensions (mm)			Water flow (q) (cc/pore cm H_2O sec) (× 10^{-4})	Water flow through gills (Nq) (cc/sec cm H_2O)	Gill area	
				d	$\frac{1}{2}b$	l_{max}			mm² × 10^4	mm²/g
0·332	252	0·0356	0·706	0·025	0·0588	0·133	0·288	0·204	0·024	724
2·71	1,285	0·0346	3·72	0·024	0·0686	0·269	1·472	5·48	0·31	775
25·98	3,409	0·0356	9·56	0·025	0·0808	0·52	0·984	9·40	1·13	436
41·1	4,121	0·0368	11·21	0·026	0·0882	0·63	1·032	11·58	1·69	411
115·72	6,540	0·042	15·56	0·031	0·0882	0·80	1·388	21·6	3·60	311
189·25	8,290	0·0453	18·28	0·032	0·1250	0·83	2·058	37·6	5·08	268·5
288·6	10,010	0·049	20·42	0·034	0·1323	0·92	2·410	49·2	7·10	246
452·0	11,890	0·049	24·22	0·034	0·1617	1·06	2·50	60·6	10·08	239
618·2	14,870	0·056	26·57	0·038	0·1764	1·24	3·262	86·8	14·29	220·5
837·5	16,210	0·0535	30·27	0·036	0·1764	1·46	2·342	71·0	18·89	225

extent interdependent, and their respective values will vary according to the conditions under which diffusion takes place. We may consider two such conditions (Fig. 3.10): (1) both water and blood are streaming laminarly over the gill surface; (2) there is complete turbulence in both fluids. The real situation will be between these extremes, but, considering the geometry of the gills, it seems reasonable to assume that the pattern of streaming, at least in water, will be nearly laminar.

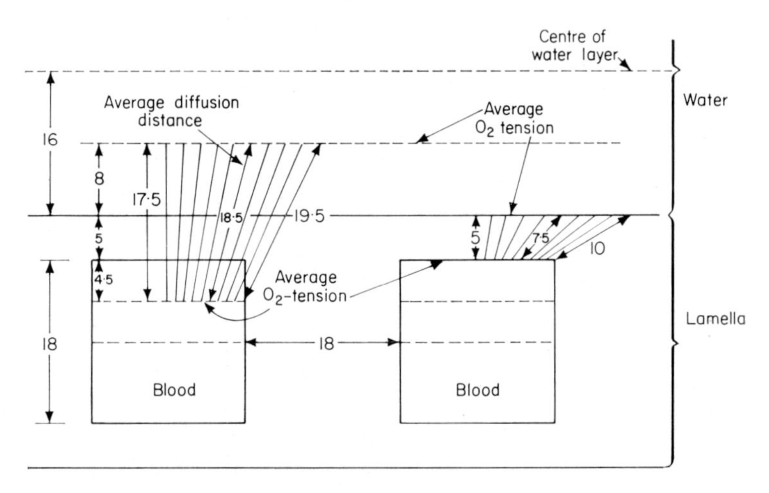

Fig. 3.10. Diffusion distance under two extreme diffusion conditions in the gills. Left: no convection in either water or blood. Right: complete convection in both. The distances are given in μ and are based on measurements of fresh gills (from Steen and Kruysse, 1964).

We shall take the gills of the common eel as an illustrative example of the uncertainties involved when we attempt to calculate the O_2-uptake (Steen and Kruysse, 1964). In a 500 g eel the total lamellar area was 1060 cm^2. The dimensions of the lamellae are seen in Fig. 3.10. The P_{O_2} of affluent and effluent water and of arterial and venous blood was measured and the results are summarized in Fig. 3.11.

The O_2-uptake for the two flow conditions can be calculated as follows:

Condition 1. Assuming that the P_{O_2} decreases linearly from water to blood, the average P_{O_2} of the water is found in a layer midway between the centre of the water layer and the tissue surface, while the average blood P_{O_2} is found midway between the centre of the blood space and the inner surface of the lamellae.

As shown in Fig. 3.10 this gives a perpendicular diffusion distance of 8 μ through the water, 5 μ through tissue and 4·5 through blood. Owing to the fact that water covers the entire lamellar surface, whereas blood covers only about half of the internal lamellar surface (Fig. 3.8), the real diffusion distances will be somewhat larger. Instead of using increased distances for each medium, and different diffusion constants, the calculation is simplified by using a somewhat increased total distance together with a compromised diffusion constant. Since the total perpendicular distance between water and blood of average tension is 17·5 μ and the longest oblique distance between the layers is 19·5 μ, 18·5 μ can be used as the average diffusion distance.

Krogh (1941) gives a diffusion constant of 0·14 ml $O_2 \times \mu/cm^2 \times atm \times min$ for muscle and 0·34 for water. The diffusion properties of gill tissue probably resemble those of muscle. Less accurately, blood may be said to have diffusion properties similar to those of water. Furthermore, gill tissue comprises about one-quarter the total diffusion distance. It therefore seems reasonable to use 0·29 as an average diffusion constant.

Using these values in the usual diffusion formula, we obtain an oxygen uptake of:

$$V = 0{\cdot}29 \frac{1060 \times 0{\cdot}13 \times 60}{18{\cdot}5} \text{ ml/h} = 130 \text{ ml/h}$$

Condition 2. The average gradient will cause diffusion only across the tissue barrier. This is 5 μ thick, but owing to the fact that blood covers only one-half of the inner surface, the mean distance will be longer. Fig. 3.10 shows that 7·5 μ is a reasonable average value. Since diffusion occurs only through tissue, it is probably most correct to use the diffusion constant for muscle tissue. This calculation gives an O_2-uptake of

$$V = 0{\cdot}14 \frac{1060 \times 0{\cdot}13 \times 60}{7{\cdot}5} = 154 \text{ ml/h}$$

The calculated O_2-uptake is thus very similar in the two conditions. In order to simplify the further discussion, 140 ml O_2/h is chosen as the calculated oxygen-uptake. Van Dam (1938) recorded an oxygen-uptake of 11·6 ml/h for a resting 450 g eel at 18°C, while Saunders (1962) found that swimming caused a 5- to 7-fold increase in oxygen-uptake in the bullhead, carp and sucker. Van Dam's value is about one-twelfth of the calculated value presented above. The calculated oxygen-uptake corresponds well, therefore, with the value which one would expect to find in a swimming eel of this size. Using the same calculation procedure, Saunders (1962) and Hughes (1966) also found oxygen-uptake values which corresponded to the maximum oxygen-uptake measured in the same fishes.

As mentioned earlier the passage time of water and blood through the gill sieve is of major importance for the efficiency, yet it is not included in the formulae for O_2-uptake. The passage time is reflected, however, in the gradient of gas tension and this parameter is included. The longer the passage time, the smaller the gradient and the higher the degree of extraction.

It is further implicit in the architecture of the gills that the same degree of utilization can be achieved by rapid flow through narrow pores as by slower flow through wider pores.

(2) In vivo *measurements of gas exchange*. The use of modern electronic equipment has resulted in a rather complete picture of gas exchange in fishes. Let us, however, start with the classical study of gas exchange in fishes by van Dam (1938). He demonstrated that the eel may extract up to 85% of the O_2 present in the inspired water. As has been mentioned earlier the limit of extraction is set by the venous P_{O_2}. In resting eels Steen and Kruysse (1964) found an average venous P_{O_2} of 15 mm Hg at an inspiratory P_{O_2} of 150 mm Hg. Thus the maximum extraction is

$$\frac{150-15}{150} \times 100\ \% = 90\ \%$$

This shows that under certain conditions the respiratory system of the eel is able to achieve almost complete O_2-exchange. Other investigations, both on teleosts (Saunders, 1962) and on elasmobranchs, reveal a lower degree of extraction, but still a venous P_{O_2} close to zero (Piiper and Schumann, 1967; Lenfant and Johansen, 1966; Piiper and Baumgarten-Schumann, 1968). In some cases (Steen and Kruysse, 1964; Piiper and Schumann, 1967) the P_{O_2} of arterial blood is higher than in effluent water. This is satisfactorily explained as a consequence of efficient exchange in a counter-current system. Other explanations have been suggested (Piiper and Schumann, 1967) but have later been rejected (Piiper and Baumgarten-Schumann, 1968).

In elasmobranchs which have been studied in detail by Lenfant and Johansen (1966) and by Piiper and Schumann (1966) the arterial P_{O_2} is often 100–120 mm Hg at inspired P_{O_2} about 150. In some cases the effluent P_{O_2} is 50–80 mm Hg. This signals a typical counter-current exchange. In several experiments, however, the effluent P_{O_2} was higher than the arterial P_{O_2}. Rather than take this as an indication against counter-current flow, I prefer to explain it by variations of blood and water distribution in the gill apparatus.

A comparison between P_{O_2} of water and blood in passive and active eels reveals a curious feature (Steen and Kruysse, 1964). Resting fishes extract about 85% of the O_2: thus at inflowing P_{O_2} of 150 mm Hg the effluent P_{O_2} is about 20 mm Hg. This shows that almost all the water must have passed circulated gill lamellae. Still the arterial P_{O_2} of resting fishes is often as low as 50 mm Hg. Swimming fishes extract only 10%, thus the effluent P_{O_2} is 125 mm Hg. Under such conditions the arterial P_{O_2} is *higher* than in the resting fish (Fig. 3.11). These findings may be explained as follows: During rest all parts of the irrigated lamellae are also circulated giving high extraction. Not all the blood passes lamellae; instead, some takes the non-respiratory shunts. The "arterial P_{O_2}" is therefore the P_{O_2} of lamellar high P_{O_2}

blood mixed with low P_{O_2} shunt blood. During activity an increasing proportion of the blood flow takes the respiratory path, thus giving a higher arterial P_{O_2}. The low O_2-extraction in active fishes may result from an increased part of the water not passed between lamellae.

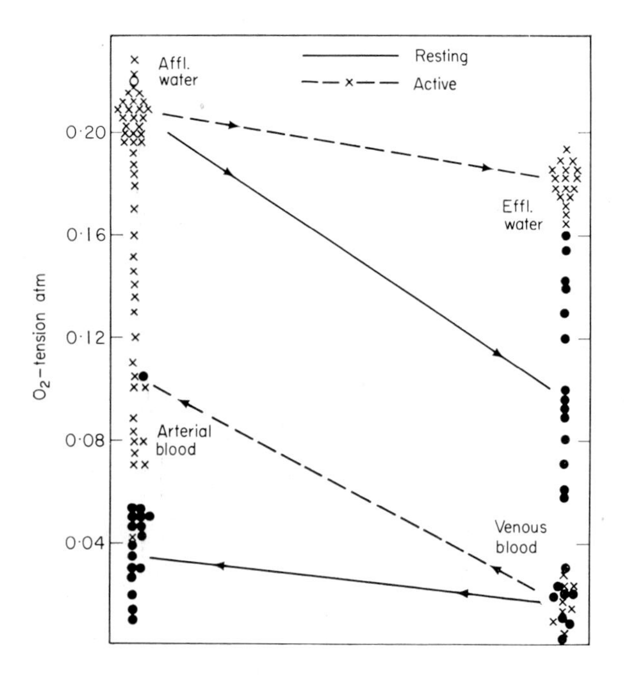

Fig. 3.11. Oxygen tension of blood and water before and after passing the gill membrane. The length of the abscissa represents the depth of the lamellae (from Steen and Kruysse, 1964).

This explanation is not experimentally verified; to do so would require rather clever experimentation. Aside from the observations mentioned, it is based on the author's belief that gas exchange between lamellar blood and intralamellar water is very efficient under any circumstances. The apparently poor exchange signalled by low arterial P_{O_2} is due to admixture to the samples we obtain by "non-respiratory" blood. This explanation also entails an optimal compromise between exchange of respiratory and non-respiratory substances as will be discussed later in this chapter.

The O_2-uptake of swimming fishes is 5–10 times above the resting level (Saunders, 1962). This is accomplished by increased cardiac output, increased ventilation and increased A–V O_2 content difference. While in higher animals an increased A–V difference is accomplished by decreased venous P_{O_2}, it appears that in fishes it is also obtained by increased arterial P_{O_2} (Steen and

Kruysse, 1964). This is possible since resting fishes often show incomplete arterial saturation.

3. ADAPTATIONS OF THE GILL FUNCTION TO VARIOUS OPERATIONAL CONDITIONS

There exists considerable variation in each respiratory parameter within fishes. In some cases it is possible to make generalizations concerning the operational conditions of a species and its respiratory characteristics.

a. Activity

It is not surprising to find a good correlation between the respiratory system and the bodily activity of a fish. Hall and Gray (1929) and Root (1931) found that active fishes, like the mackerels, have more Hb per ml blood, i.e. a higher O_2-capacity, than more lethargic species like the eel and the toadfish (Table 3.3). Gray (1954), Hughes (1966) and Steen and Berg (1966) showed that active fishes also had a larger gill area per gm body weight (Table 3.3). Steen and Berg (1966) found that active fishes also have a shorter diffusion distance between blood and water (Table 3.4). The O_2-affinity does not, on the other hand, show any clear correlation with activity. This may be due partly to the lack of P_{50} values at *in vivo* conditions and partly to the fact that P_{50} correlates with the environmental P_{50}. Apart

TABLE 3.3
Blood O_2-capacity, total gill area and lamellar frequency in some fishes (from Hall and Gray, 1929; Gray, 1954; Hughes, 1966).

Species	O_2-capacity vol. %	Gill area mm²/g body weight	Lamellae per mm	
Bonito	18·0	595	—	
Bulls-eye mackerel	17·5	—	—	
Common mackerel	17·0	1158	29	Very active
Menhaden	16·2	1773	27	
Cunner	10·9		—	
Butterfish	10·7	598	32	
Scup	10·0	506	25	Active
Sea-robin	9·3	360	20	
Rudderfish	8·6	506	—	
Puffer	8·5	470	18	
Eel	8·0	302	18	
Silver-hake	7·7		—	
Goosefish	5·7	196		
Toadfish	5·3	200	11	Sluggish
Sand-dab	4·6	188		

from this parameter, however, we can conclude that the respiratory mechanism of active fishes is better designed for gas exchange than that of less active ones. Teleologically this is somewhat surprising as it is hard to see what can be gained by, for example, a thick blood–water barrier. The bullhead has a barrier more than 10 times thicker than the mackerel. It could have saved nine-tenths of its gill area instead. The reason for this apparently unreasonable situation may simply be that there is nothing to be gained for this fish by making the gills a more compact organ.

In a few cases of large, fast-swimming pelagic fishes, the rows of filaments are contained between hard perforated plates (Muir and Kendall, 1968). The perforation in the blue marlin corresponds to at most 5% of the total side surface of the filaments. These fishes are distinguished by a high swimming velocity and therefore by a high water velocity through the gills. The perforated cover will necessarily restrict the velocity of water flow between the lamellae. This may serve to shield the delicate lamellae (8–10 μ thick) against damaging mechanical turbulence and possibly to maintain a water flow between them which is in accordance with the blood flow. It is possible that the extensive system of gill rakers found in many other fishes and the cilia of molluscan gills may serve a similar function, i.e. to check the flow of water in direct contact with the respiratory surfaces.

b. Environmental Gas Composition

The gill dimensions of many fishes from normal well-aerated waters are shown in Tables 3.3 and 3.4. The gills of about 50 species of fishes from the Amazon river, which has a P_{O_2} of 50–70 mm, were studied by Junquiera, Steen and Tinoco (not published). Their gills were distinguished by a very short blood-to-water distance, less than 1 μ in 45 out of 50 species. Another typical feature was that the distance between adjacent gill lamellae was usually small compared to the width of the lamellae.

Thus while fishes from normoxic water have 2–5 times wider water space than lamellar width, the Amazon fishes only rarely had more than 2 times thicker water than blood space. These structural features entail that the water-to-blood diffusion distance is smaller in fishes from O_2-poor water than in fishes from well-aerated water. This is a rational compensation for the lower gas gradient across the gills of fishes in this habitat.

The properties of the blood also parallel the composition of the habitat. Fish (1956) studied the blood of six species of fishes which inhabited different habitats in Uganda. Inspection of Fig. 3.12, which is from his work, illustrates this point. *Lates* live only in well-aerated water and die in large numbers when occasionally exposed to low ambient P_{O_2} or high P_{CO_2}. At the other extreme is *Bagrus* which inhabit deep and very O_2-poor layers. Willmer (1934) found much the same pattern: fishes from well-aerated water have blood with a low O_2-affinity which is considerably increased by lowered pH.

TABLE 3.4

Gill dimensions in some active and sluggish fishes (from Steen and Berg, 1966).

Species	Thickness of lamellae (μ)	Lamellae (per mm)	Distance between lamellae (μ)	Distance between blood and water (μ)		Type of epithelium
Icefish (*C. aceratus*)	35	8	75	6		Cubical, nuclei in diffusion path
Bullhead	25	14	45	10	Sluggish fishes	Cubical, nuclei in diffusion path
Eel	26	17	30	6		Flat, no nuclei in diffusion path
Sea scorpion	15	14	55	3		Flat, no nuclei in diffusion path
Flounder	10	14	70	2		Flat, no nuclei in diffusion path
N. Tessellata	20	17·5	35	2		Flat, no nuclei in diffusion path
Icefish (*D. esox*)	10	18	40	1		Flat, no nuclei in diffusion path
Trout (400 g)	12	23	35	3	Active fishes	Flat, no nuclei in diffusion path
Roach	12	27	25	2		Flat, no nuclei in diffusion path
Coalfish	7	21	40	<1		Extremely flat
Perch	10	31	25	<1		Extremely flat
Herring	7	32	20	<1		Extremely flat
Mackerel	5	32	25	<1		Extremely flat

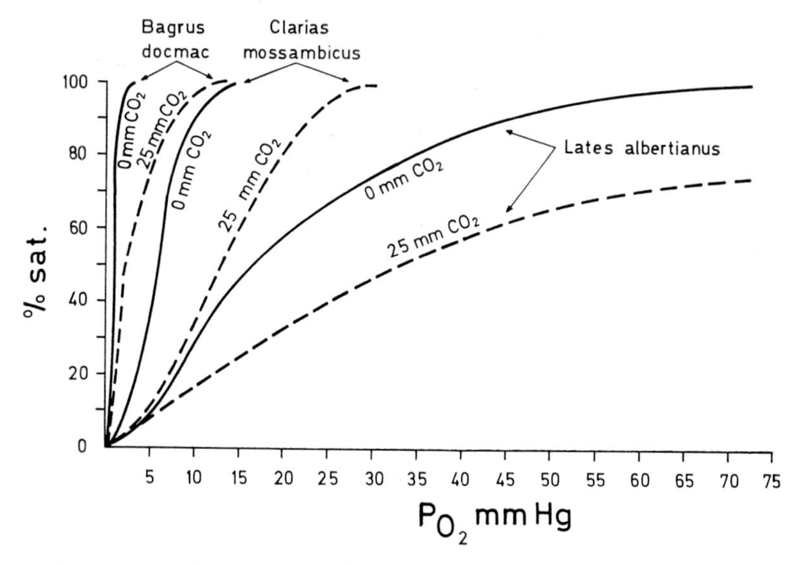

Fig. 3.12. Oxygen-equilibrium curves for blood of three teleosts at $P_{CO_2} = 0$ and 25 mm Hg (redrawn from Fish, 1956).

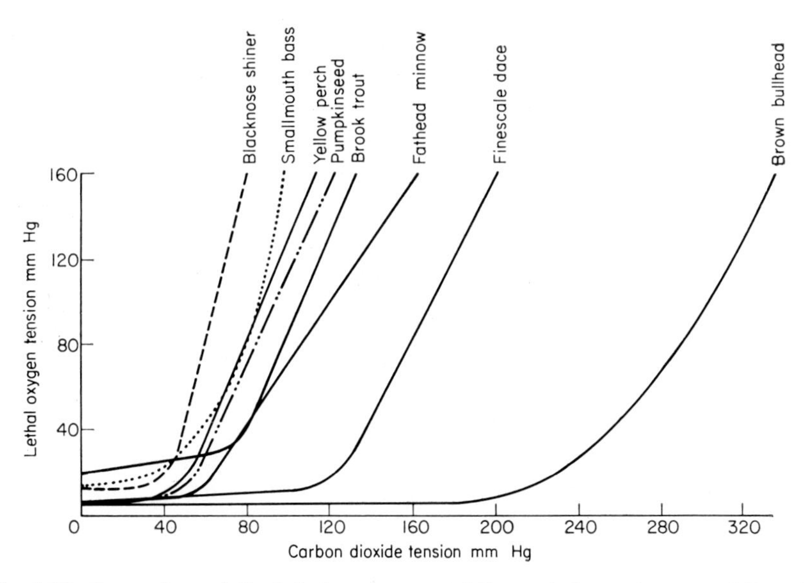

Fig. 3.13. Comparison of the lethal environmental P_{O_2} and the environmental P_{CO_2} for various fishes (from Black *et al.*, 1954).

Fishes from O_2-poor and acid waters have blood with high O_2-affinity which is less sensitive to pH reduction. Also, reduced O_2-capacity as a result of acidification (Root-effect) seems to occur only in blood from fishes of well-aerated waters. An example is illustrated in Fig. 3.12.

A high O_2-affinity of the blood is paralleled by an ability to utilize a larger proportion of the ambient O_2 and thus survive hypoxic conditions better than fishes with blood of lower O_2-affinity (Hall, 1930; Black *et al.*, 1954).

The same authors also demonstrated that the "lethal" P_{O_2} increased with increasing P_{CO_2} (Fig. 3.13). The fact that the fishes suffocated at high P_{CO_2} even at an ambient P_{O_2} of 160 mm Hg could indicate that they had blood with a Root-effect, i.e. that the lowered pH caused reduction in O_2-capacity (Ch. 1). However, comparing these deaths with the survival of fishes with Hb inactivated by CO (Nicloux, 1923) it is most likely that the main cause of death was acidosis rather than hypoxia. It may be significant that none of the fishes that we investigated from the Amazon river, which is very acid and has a low P_{O_2} exhibited a Root-effect (personal observation).

The relationship between the O_2-affinity and the magnitude of the Bohr-effect is not a simple one. Black (1940) found that a standard decrease in pH caused the P_{50} to increase more, the higher the original P_{50} (at approximately arterial pH). This is evident in Fig. 3.12. However, the relative change may often be larger in bloods of low O_2-affinity. Thus in the carp, with a P_{50} of 3 mm at low P_{CO_2}, an increase of 20 mm P_{CO_2} does not cause the P_{50} to increase above 10 mm, while in the common sucker, which has a P_{50} of 10 mm, a similar increase in P_{CO_2} causes the P_{50} to increase above 60 mm.

Willmer (1934) described a good correlation between the magnitude of the Bohr-shift and the habitat in a sample of fresh-water fishes which had the same P_{50} at low P_{CO_2}. The paku which lives in well-aerated water showed an increase of P_{50} from 10 to 50 mm Hg upon addition of 25 mm CO_2. Similar hypercarbia caused increase in P_{50} in the blood of two other species from less well-aerated water from about 10 to 20–30 mm Hg, while it caused almost no reduction in the O_2-capacity of the blood from fish inhabiting acid marsh water (Fig. 3.14).

It is important to be aware of the different physiological consequences imposed on a fish that has blood with a Bohr-effect only, compared to those with a Root-effect in addition. In the former case full O_2-saturation is attainable, independent of blood pH, as long as the P_{O_2} approaches 0·2 atm. Thus, a Bohr-effect affects gas exchange across the gills only at low P_{O_2}. In the tissues the Bohr-effect will facilitate gas exchange. The Root-shift introduces a more serious consequence of low blood pH, since regardless of the P_{O_2} the blood will remain unsaturated. Its effect thus resembles an anaemia, for example, partial CO poisoning.

We should also appreciate the possible role of the rate of the pH-induced changes in the blood. When eel erythrocytes are acidified in *in vitro* experiments, O_2 is released with a half time of about 50 msec at 20°C. When the process is reversed, that is when the pH of red cell suspensions is raised, O_2 is absorbed with a half time of several seconds (Forster and Steen, 1969).

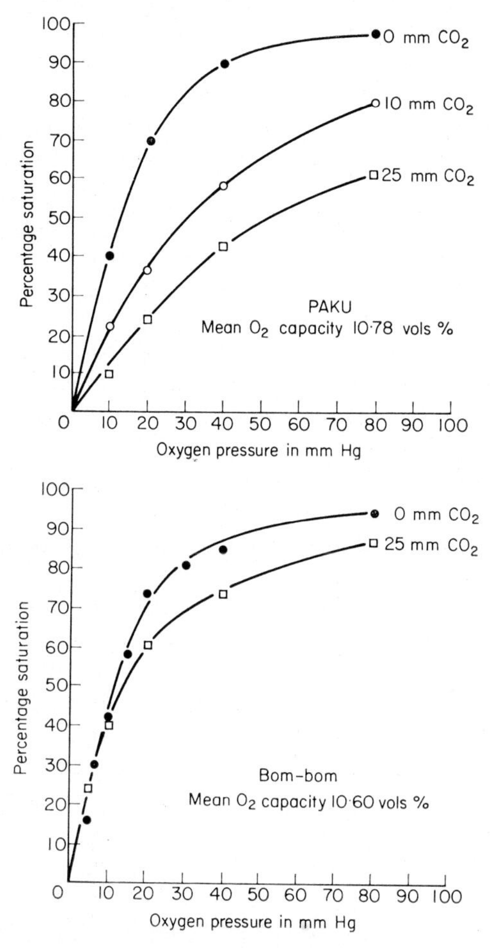

Fig. 3.14. Above: oxygen-equilibrium curves for blood from the bom-bom which inhabits acid water. Below: oxygen-equilibrium curves for the paku which inhabits well-aerated water. The O_2-affinity of the acid water inhabitant is far less P_{CO_2}-sensitive than that of the "normal" water inhabitant (from Willmer, 1934).

Although no experiments have yet been conducted to see if this has consequences for gas exchange, there are several possible effects of the kinetics of these processes. Such information is needed before we can appreciate the physiological importance of the Bohr- and Root-effect more appropriately.

c. Icefishes — Importance of Haemoglobin

We are accustomed to think that Hb is a vital component of blood in all vertebrates. This is evidently not so in all fishes. Nicloux (1923) found that eel, pike and carp will survive for hours in water equilibrated with air containing enough CO to make Hb completely inefficient as an O_2-carrier. Anthony (1961) showed that goldfish display normal activity in water equilibrated with 80% CO and 20% O_2, provided the temperature is below 20°C. Even more striking is the unintentional experiment of Norwegian fishermen in which a fishing hook in the dorsal vein rendered an eel completely anaemic while its behaviour, judged by eel connoisseurs (Steen and Berg, 1966), was indistinguishable from that of normal eels with a blood O_2-capacity 10–20 times higher. The most striking demonstration, however, is given by the Hb-free Antarctic fishes whose blood has an O_2-capacity no higher than that of sea water (Ruud, 1954).

The gill anatomy of these so-called icefishes has been investigated by Hughes (1966), Steen and Berg (1966) and Jakubowski *et al.* (1969). One of these icefishes, *Chaenocephalus aceratus*, is a slow bottom-dweller weighing 0·5–1 kg. The gills of this specimen are unusually coarse with low lamellar frequency and wide blood lacunae in the lamellae (Table 3.4). It has all the structural features which signal a low resistance system, both ventilatory and circulatory. Another species *Chamsocephalus esox* is pelagic and weighs about 100 g. Its gills do not differ structurally from fishes that possess Hb, but are otherwise comparable in size and habitat. The same holds true for a third species of icefish *Chaenichthys rugosus* (Jakubowski *et al.*, 1969).

These authors also measured capillarization of the gills and of the body surface of their icefish and of Hb-possessing fishes. In the latter the gills have 3–30 times as many capillaries, expressed per g body weight, in the gills as on the body surface. In the icefish areas other than the gills had 3 times as high capillary density as the gills. The respiratory significance of these data is uncertain since the blood-to-water distance is not known for the skin capillaries. Hemmingsen and Douglas (1970) found that cutaneous gas exchange made up about 40% of the total O_2-uptake. It is puzzling that cutaneous O_2-uptake should be high in icefishes. One would think that an increased gill area would do just as well.

The respiratory situation of icefishes has recently been investigated by Hemmingsen and Douglas (1970) who also reviewed the literature on the subject. The total resting O_2-uptake of these fishes appears to be one-half to one-third of that found for Hb-possessing fishes from the same habitat. The O_2-uptake was unaffected by the ambient P_{O_2} as long as this was above about 50 mm Hg. The icefishes therefore exhibit a considerable respiratory capacity despite the fact that their blood has an O_2-capacity of only 0·7 volume per cent compared to about 8 volume per cent for the Hb-possessing fishes.

Today we have at least a qualitative explanation of how the low blood O_2-capacity is compensated. Hemmingsen and Douglas (1970) showed that the blood volume of icefishes made up 7·5% of the body weight as compared to 2–3% in normal fishes. This latter observation suggests, together with the wide blood capillaries of their gills and myoglobin-containing heart, that an increased cardiac output plays an essential role in obtaining the high O_2-uptake. The lack of red cells also decreases the viscosity of the blood. This allows an increased blood flow velocity compared to normal fishes at the same energy of propulsion.

Hemmingsen and Douglas also measured P_{O_2} of inspired and expired water and of venous blood. Venous P_{O_2} averaged 90 mm Hg when the fish was in water of P_{O_2} above 160 mm Hg and the expiratory P_{O_2} was 10–25 mm Hg lower. These data indicate that O_2-exchange both across the gills and in the tissues most likely occurs at a large ΔP_{O_2}.

d. Defensive Adaptations to Non-respiratory Exchange

In Chapter 2 it was mentioned that one of the problems inherent in aquatic respiration was the exchange of non-respiratory substances, primarily ions. The respiratory mechanism of fishes shows features which indicate that a certain degree of defence against such exchange is incorporated into their respiratory mechanism.

As described in section D.1 of this chapter, fish gills appear to have a double circulatory path: one respiratory through the lamellae, and one non-respiratory through the interior of the filament. Steen and Kruysse (1964) showed that in excised filaments blood tended to take the respiratory path when adrenaline had been added to the bathing solution, but the non-respiratory path when acetylcholine was present. The functional significance of this double circulation was indicated by the observation that intravascular injection of adrenaline increased the arterial P_{O_2}. Baumgarten *et al.* (1968) showed that the Na^+ outflux through the gills increased during activity, and increased further when nor-adrenaline was added to the water. This probably means that the increased ion loss is due to an increased blood flow through the respiratory path of the gills. The vascular control of blood flow to the two paths may therefore serve as a kind of preventive osmoregulation. The presence of a respiratory pigment increases the gas-to-ion exchange ratio since it increases the gas capacity, but not the ion capacity. An optimal relationship between the two exchange processes is further enhanced if the rate of blood flow is regulated so that the gas gradient is as large as possible. This would be accomplished if the P_{O_2} of blood leaving the lamellae corresponded to the upper portion of the steep part of the dissociation curve. In most cases this would imply that this P_{O_2} should be well below the P_{O_2} of the water entering the gill apparatus (affluent water).

Chapter 4

TRANSITIONAL BREATHING

A. The Transition from Aquatic to Aerial Breathing

Most water-breathing animals are able to exchange gases also with air, but in most cases in amounts insufficient for sustained survival. In several aquatic habitats, we find animals that are typically adapted in various ways to breath air directly. In the tidal zone, and in rivers and swampy lakes of low O_2-content, some of which subjected to periodic droughts, we find animals showing a dual mode of breathing. Inhabitants of these habitats may show anatomical and physiological characteristics which made air breathing and later terrestrial life possible. Knowledge of these animals may therefore throw light upon the evolution of air breathers.

As discussed in Chapter 2, air is the most favourable respiratory environment. In line with this we find that a number of water breathers have accessory organs for air breathing, whereas only a few air breathers possess accessory organs for water breathing. Secondary adaptation to an aquatic mode of life has instead involved either bubble or plastron breathing of insects (Ch. 5), or a diving syndrome as in reptiles, birds and mammals (Ch. 7).

There are only a few cases where the same organ serves gas exchange with both water and air. In water the gill lamellae are supported and kept apart by the medium itself, while in air they collapse like sea weeds on the beach and the respiratory area is drastically reduced. Excluding this mechanical aspect, however, gills appear to be fully suited for use in air. This is attested to by the fact that many animals with more rigid lamellae respire with air.

Thus even though gills may be useful in air, lungs are better. Lungs are typically made so that they are kept open in a stream of air: either air capillaries, as in bird lungs, or dead-end alveoli as in other lungs. Other air cavities are kept open by their own mechanical rigidity.

It is often stated that gills are poorly fitted for use in air owing to concomitant water loss by evaporation. This is an oversimplification since it applies only to gills that are freely exposed to the atmosphere and very few are. Respiratory water loss from organs enclosed in a cavity is a function of the rate of ventilation. The problem of desiccation is on the other hand reflected by the fact that gas exchange across the general body surface does not play any important role in air-breathing animals, while it is often significant among water breathers.

Physiological studies of dual breathers have mostly been concerned with adult animals. Unfortunately little information exists on the respiratory mechanisms of animals whose ontogenetic development mimics a phylogenetical trend in evolution of air breathing. Some of this information is discussed in connection with the respiratory mechanisms of bird eggs. Some of the information on amphibia during their development from aquabreathing tadpoles to air-breathing adults is discussed at the end of this chapter.

1. REACTIONS TO HYPOXIC WATER

Almost any water breather will react when the environmental P_{O_2} falls below a certain limit. The reaction varies among different animals, but the most common, and the interesting one from our point of view, is that the animals seek the surface. There they either irrigate their gills with water from the upper O_2-richer layer or they switch to aerial respiration. The first type of reaction was described by Wallengren (1914) for dragonfly larvae when the P_{O_2} fell below 55 mm Hg. A similar reaction is found in the fish *Leuciscus erythrophthalamus* when the P_{O_2} falls below 15 mm Hg (Winterstein, 1908). During the 1967 Alpha Helix Expedition to the Amazon basin, this reaction was illustrated to us by a fresh-water ray (*Paratrygon* sp.). It was placed in an aquarium and took its place on the bottom as rays are expected to do. However, when the P_{O_2} of the water approached 15 mm Hg it rose to the surface and let water pass in shallow streams from all points of the circumference to the central respiratory opening. In this way it utilized the more O_2-rich surface water and preoxygenated it further by spreading it in a thin film over its flat dorsal surface.

Another reaction to hypoxic water is to seek the surface and gulp air to supplement gas exchange. This is found in many teleosts that show no anatomical adaptations for air breathing (goldfishes, eels), but in most cases such fishes do show typical structural adaptations. Such animals are termed dual breathers and representatives are found both among invertebrates and among vertebrates.

2. INVERTEBRATE DUAL BREATHERS

In the intertidal zone there are some interesting examples of invertebrates which exchange gases with water during high tide and directly with air during low tide (Krogh, 1941). The opistobranch *Ancula* and several shore crabs (*Grapsus*, *Carcinus*) all of which have sufficiently rigid gills to support respiration in air, are examples of bimodal breathing. A most interesting case is presented by the pagurid crustacean *Birgus latro*. This animal has a combined gill-lung chamber, where the gills occupy the lower half and the lungs the roof (Fig. 4.1). Evidently the animal exchanges gases as well in air, as in water, but must return to water periodically, not to breath, but to moisten the respiratory surfaces (Semper, 1878).

Unfortunately the physiological response of these animals to water and air exposure has not been investigated. However, the principle problems of their situation appear similar to those of air-breathing fishes whose physiology has been extensively investigated (Johansen, 1970).

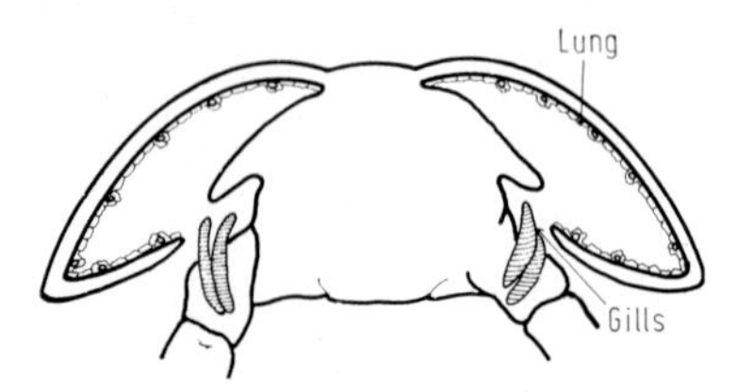

Fig. 4.1. Schematic drawing of the combined gill-lung chamber of the crustacean *Birgus latro* (simplified from Semper, 1878).

3. AIR-BREATHING FISHES

There are a few fishes, like the common eel *Aguilla vulgaris*, which can respire in air, without showing any marked anatomical differences compared to the typical teleost gills. The gills of the eel are typical for fairly active teleosts. And, as the folklore goes, it can eel its way through the dew-wet morning grass. While in air, the eel uses its gills for aerial respiration. It takes a breath and keeps the gas in the gill cavity until the P_{O_2} has reached about 100 mm Hg. Then the opercula open and the gas is expelled before a new volume of air is taken in. At a temperature of 7°C such aerial gill respiration, together with gas exchange across the skin, seems to cover the metabolic needs, since no lactic acid is accumulated (Berg and Steen, 1965). At higher temperatures, however, the arterial O_2-content is below normal and the eel develops metabolic acidosis and an O_2-debt.

The majority of air-breathing fishes show structural adaptations for gas exchange with air.

At least four principally different types of such structures can be distinguished:

a. *Mouth breathers* with increased vascularization of mouth and pharynx.
b. *Intestinal breathers* with increased vascularization of the gastro-intestinal tract.
c. *Bladder breathers* with a swimbladder supplied with post-branchial blood.
d. *Lung breathers* with a lung supplied with pre-branchial blood.

a. Mouth Breathers

In most cases the presence of accessory structures for air breathing is accompanied by a reduction of the gills. However, there are a few fishes where both efficient gills and air-breathing organs are present. A good example is the South American *Symbranchus marmoratus*. Their gills are unique among teleosts in that each gill arch carries only one row of filaments, but every other filament extends to opposite sides (personal observation). This provides sufficient spacing so that they do not stick together during air exposure. However, the lamellar frequency and dimensions are as in regular water-breathing teleosts and the lamellae stick together in air. In addition *Symbranchus* has a well-developed organ for air breathing in that the roof of the mouth is richly vascularized (Fig. 4.2). This fish therefore possesses specialized organs for respiration in both media.

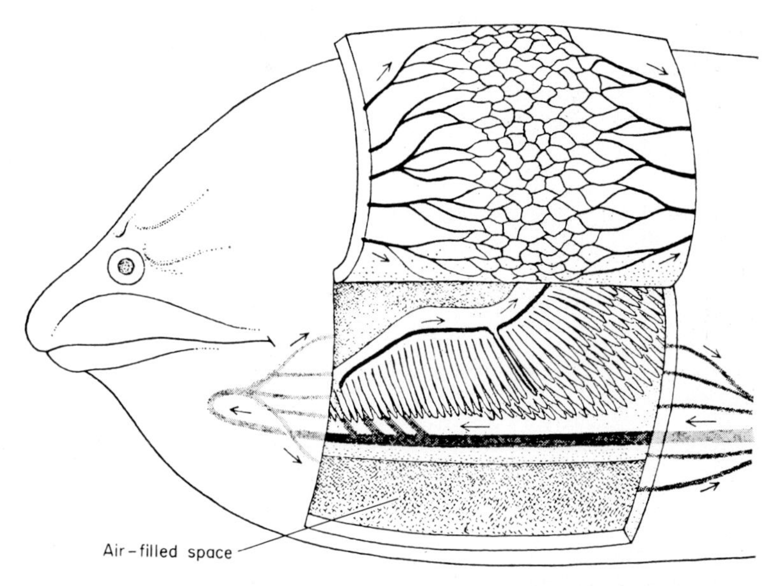

Fig. 4.2. Schematic drawing of the gas exchange surfaces of *Symbranchus*. Gills are modified and are used both in air and in water. In addition the gill covers are richly vascularized (flap) (from Johansen, 1968).

The accessory gas exchange surface of *Symbranchus* is supplied by venous blood, parallel with the gills and drained into the systemic arterial system (Fig. 4.6C). Circulation of the two respiratory organs is therefore arranged in parallel.

Johansen (1966) examined gas exchange in *Symbranchus* alternately exposed to air and water. During air exposure it filled the mouth cavity with air and expelled it when the P_{O_2} reached a certain level (Fig. 4.3). Both gills and mouth cavity take part in this gas exchange. As shown in Fig. 4.4 the arterial P_{O_2} is much higher when the fish is in air, as compared to when it is in water without access to air. However, the elimination of CO_2 is easier in water than in air. Periodical immersions may therefore appear necessary to get rid of excess CO_2 and probably also to avoid desiccation of the respiratory surfaces.

In most cases, specialized structures for air breathing are found in fishes with reduced or altered gill apparatus.

An example of gill reduction attending development of accessory respiratory surfaces for air breathing is found in the electric eel, *Electrophorus electricus*. Special attention will be given to this species since its respiratory

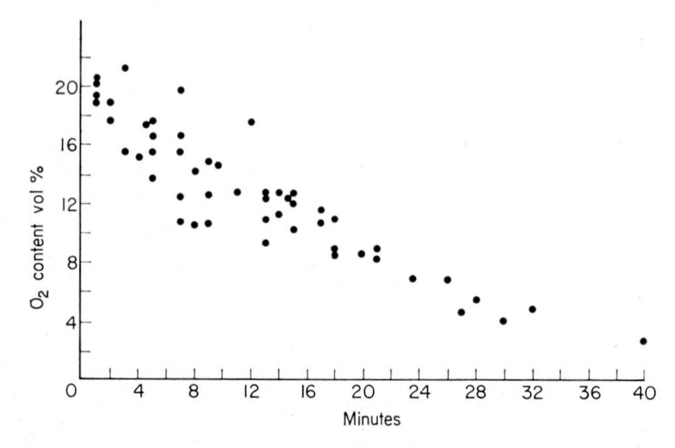

Fig. 4.3. Decline in O_2-concentration of air taken into the gill chamber of *Symbranchus* (from Johansen, 1966).

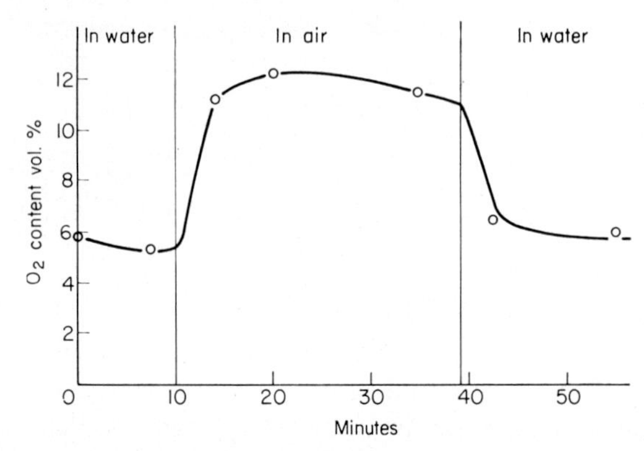

Fig. 4.4. Arterial oxygen content during water breathing interrupted by one cycle of air breathing in *Symbranchus* (from Johansen, 1966).

physiology has been well described (Johansen *et al.*, 1968). *Electrophorus* has very reduced gills; there are no proper lamellae and the exchange area is small. These gills are, however, more rigid than other gills and do not collapse in air. The mouth cavity has an elaborate system of extremely well-vascularized papillae (Fig. 4.5). Their total area, macroscopically measured, is not large—about 15% of the body surface (Johansen *et al.*, 1968)—but this is partly compensated by an unusually efficient arrangement of the capillaries. They form loops protruding above the surface, increasing the efficient surface. The distance from blood to water is below resolution of the light microscope (less than 1 μ) (personal observation). An electron-microscopical investigation of this surface would be rewarding

Electrophorus is an obligate air breather; the vestigial gills are of virtually no help for aquatic gas exchange and the geometry of the mouth papillae also prohibits such exchange. The papillae are supplied by blood from the heart in parallel with blood supply to the gills. But, whereas the gills are drained into the dorsal aorta, the papillae are drained by various veins which return the blood to the heart after mixing with deoxygenated venous blood.

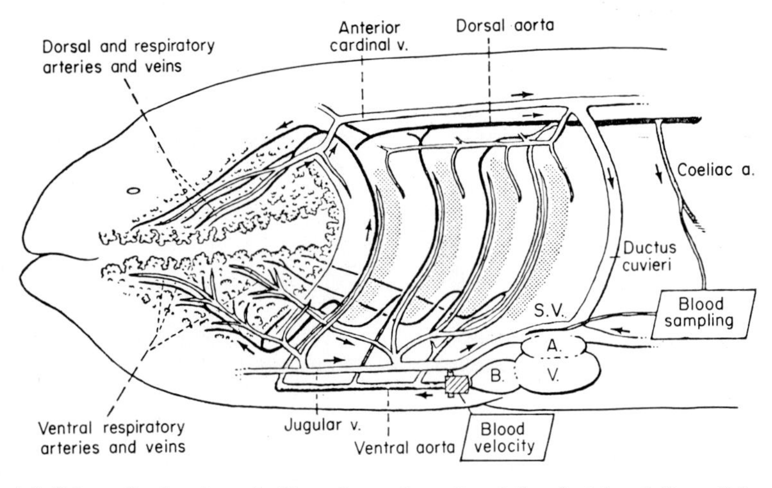

Fig. 4.5. Schematic drawing of gills and mouth cavity of the electric eel (from Johansen *et al.*, in press).

This mixed arterial blood then perfuses the non-functional gills and continues to the arterial system. In *Electrophorus* there is thus no separation of oxygenated and deoxygenated blood with regard to the arterial supply. Figs. 4.5 and 4.6B illustrate this circulatory arrangement. Johansen *et al.* (1968) showed that the ventilatory pattern of this fish did not respond to changes of the P_{O_2} or P_{CO_2} of the water, but responded promptly to such changes in the gas phase above the water. This further attests to their independence of gas exchange with water.

Johansen (1970) gives a good survey of other structural developments of the kind described for *Electrophorus*. A complicated, labyrinthine air-breathing organ of *Anabas testudineus* was described by Munshi (1968), but the principal architecture does not appear to differ from that in *Electrophorus*. In *Trichogaster trichopterus*, Schultz (1958) studied the blood–air barrier with the electron microscope and found the shortest distance recorded for respiratory barriers: a minimum value of 0·06 μ, maximum 1·2 μ and average

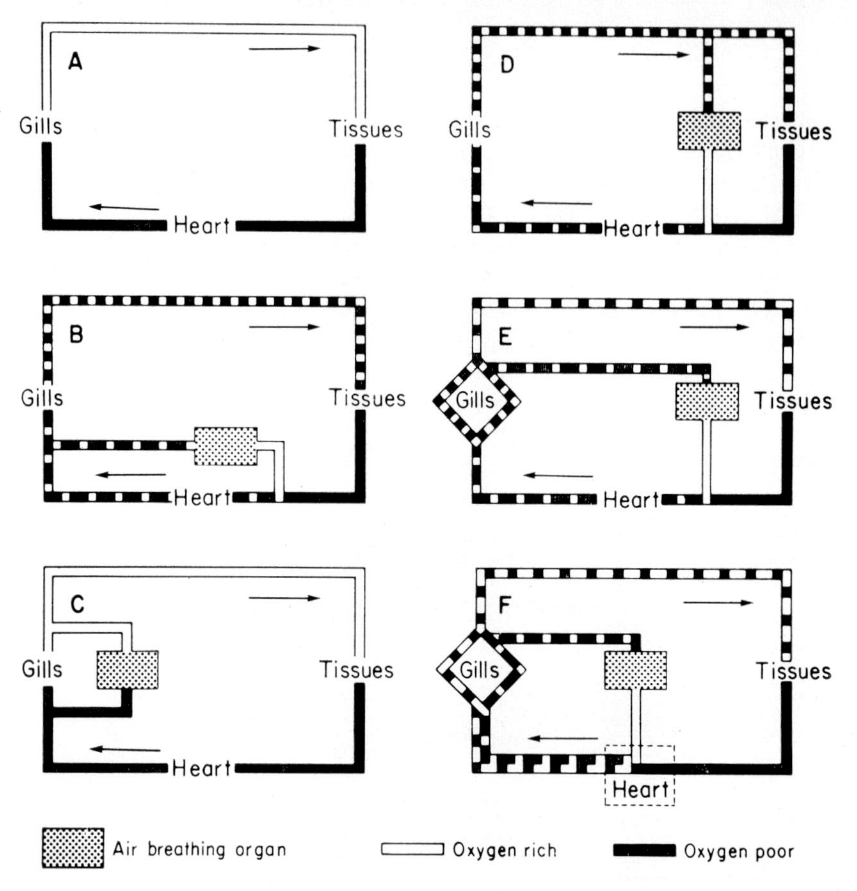

Fig. 4.6. Schematic representation of the main patterns of circulation relative to the gills and the air-breathing organs in air-breathing fishes. The amount of black and white represents only the approximate level of oxygenated and deoxygenated blood carried in the vessels.

Type A: General piscine arrangement with the branchial (gill) and systemic (tissues) vascular beds in direct series.

Type B: Air-breathing organ derived from pharyngeal and/or opercular mucosa. Afferent vessels to air-breathing organ arranged in parallel with branchial circulation and derived from afferent branchial vessels. Efferent vessels from air-breathing organ connected to systemic veins. Arrangement typical of: *Monopterus*, *Ophiocephalus*, *Electrophorus*, *Amphipnous*, *Perioptthalmus*, *Anabas*.

Type C: Gills, buccal mucosa or chambers extending from the opercular cavity serving as air-breathing organs. Afferent vessels to air-breathing organ arranged in parallel with the branchial vascular bed. Efferent circulation largely connected with the efferent branchial circulation. Systemic arterial blood highly oxygenated. *Clarias*, *Saccobranchus*, and in part, *Symbranchus* and *Hypopomus* typical of type C.

Type D: Air-breathing organ associated with the gastro-intestinal tract. Afferent circulation derived from the dorsal aorta (systemic arteries). Efferent circulation connected to systemic veins. Arterial blood mixed. Arrangement typical in: *Hoplosterum*, *Plecostomus* and *Ancistrus*.

Type E: Airbladder serving as air-breathing organ. Specialized afferent vessel derived from the most posterior epibranchial arteries. Efferent circulation connected to systemic veins. Arterial blood mixed. Arrangement typical of *Polypterus*, *Amia* and *Lepisosteus*.

Type F: Airbladder structurally advanced to resemble amphibian lung. Specialized afferent and efferent circulation in parallel with systemic arterial circulation. Partial septation of the heart allows a high degree of selective perfusion in the respiratory and systemic circuits. Arrangement seen in lungfishes and in *Protopterus* and *Lepidosiren* in particular (from Johansen, in Press).

distance 0·1–0·2 μ. Some part of the organ had exchange capillaries which lacked endothelium and basal membrane. The thin epithelium was therefore the only barrier between blood and air.

Some fishes belonging to the *Gobidae* that inhabit tidal swamps have become so adapted to aerial respiration that they cannot respire in water. In *Periophthalmus schlosseri* for example, the gill lamellae have coalesced and the gill chamber is periodically ventilated with air.

All air-breathing fishes with accessory air exchange organs in the mouth or pharynx present principally the same solution to dual breathing: an air-breathing organ as part of or in the same cavity as the gills. In some of these fishes, like *Electrophorus*, no dramatic deviation from the piscine pattern of circulation has occurred. In others, like *Symbranchus*, a definite change is apparent in that blood from the accessory organ is drained into the systemic arterial system. This difference is, however, functionally less important since the gills of *Electrophorus* are vestigial and almost non-functional, at least for O_2-exchange.

b. Intestinal Air Breathers

In some fishes part of the stomach or part of the small intestine has been modified for air breathing. When the P_{O_2} of the ambient water becomes severely low or when the fish migrates over land, it swallows air rhythmically and expires either through the mouth or the anus. Also in these fishes the blood from the accessory respiratory surface enters the venous system (Johansen, 1970) (Fig. 4.6D).

c. Bladder Breathers

In most fishes the swimbladder serves mainly as a hydrostatic organ ensuring neutral buoyancy of the fish (Ch. 10). However, it may also serve as an accessory organ for gas exchange during situations of low water P_{O_2}. The O_2-uptake from the bladder of the common eel increases when the fish is exposed to air and may constitute as much as 50% of the total O_2-uptake. However, the bladder of the eel cannot be ventilated with air; instead, its O_2-content is deposited from the arterial blood. This implies that its O_2-store can have only a temporary buffer-like function during hypoxic conditions.

Other teleosts can ventilate their swimbladder and thereby utilize it as a permanent respiratory organ. From a physiological point of view these fishes are distinguished from the lungfishes by an important feature of the circulation: the bladder is supplied by a branch from the dorsal artery and is drained into the systemic venous circulation. In lungfishes the lung is supplied by a branch from a gill artery and drained directly to the heart. The bladder breathers are therefore not able to separate venous blood from oxygenated bladder blood anatomically. In the lungfishes this important possibility exists and is utilized (see part D below).

The blood supply to the swimbladder is therefore principally the same as to the accessory respiratory organ of intestinal breathers. It is distinguished from that of mouth breathers, however, since it is supplied by blood that has passed the gills, and not by pre-branchial blood. The swimbladder may therefore be considered to be a preoxygenator for the gills. As will be discussed shortly, this may amount to the main oxygenator in situations where the P_{O_2} of the water is low.

Gas exchange in a typical bladder breather, the bowfin *Amia calva*, has recently been investigated by Johansen *et al.* (1970).

The bowfin is a very versatile fish. Its North American habitat is cold and ice-covered in winter but warm in summer. Unlike many air-breathing fishes, it is an active swimmer. Most remarkable, however, is its habit of aestivation during droughts. Fields which have been flooded at one time of the year may reveal aestivating bowfins when ploughed. This fish has well-developed gills and a vascular swimbladder which can be ventilated. Under appropriate conditions it can rely entirely on one or the other, although it normally uses both simultaneously.

Johansen *et al.* (1970) measured the gas exchange from water and air during conditions of varying temperature and ambient P_{O_2}. At 10°C *Amia* relies almost exclusively on water breathing and it visits the surface very infrequently. At 30°C it will surface at intervals of 1–2 minutes and obtain about two-thirds of its O_2-uptake via the bladder. The water-ventilated gills are, however, also the principal sites of CO_2 elimination at high temperature.

There is a gradual increase of air breathing with decreasing water P_{O_2}. At a P_{O_2} of 40–50 mm Hg bladder breathing increases sharply and becomes almost totally dominant compared to gill breathing which declines sharply at such low P_{O_2}.

Johansen *et al.* (1970) also showed that when one of the respiratory organs of *Amia* dominated, the blood was to a considerable extent shunted to bypass the exchange vessels of the other. The functional importance of this is clear. If, for example, the fish is in very O_2-poor water it might run the risk of losing O_2 picked up in the bladder via the gills.

The respiratory properties of *Amia* blood is similar to that of other dual breathers (Chs 4 and 5).

d. Lungfishes

(1) *Functional anatomy of the respiratory system.* In the lungfishes we find a most important innovation in the transition from water to air: the lung. This organ serves gas exchange primarily and in contrast to the other air-breathing organs it is a new organ and not a modification of a pre-existing one.

In spite of being phylogenetically primitive, the dipnoan lung is structurally rather specialized. Internal septa and ridges divide the air space into intercommunicating compartments which are further subdivided into richly vascularized alveolus-like pockets. Structurally these lungs are more specialized than for example the smooth surface lung of the salamander *Proteus* and resemble lungs from frogs and turtles (Fig. 4.7). The airways have more smooth musculature than that which is found in the trachea and bronchi of terrestrial vertebrates. Electron microscopical investigations have revealed a blood-air distance of about 0·5 μ (Klika and Lelek, 1967) and a microstructure similar to that in higher vertebrates.

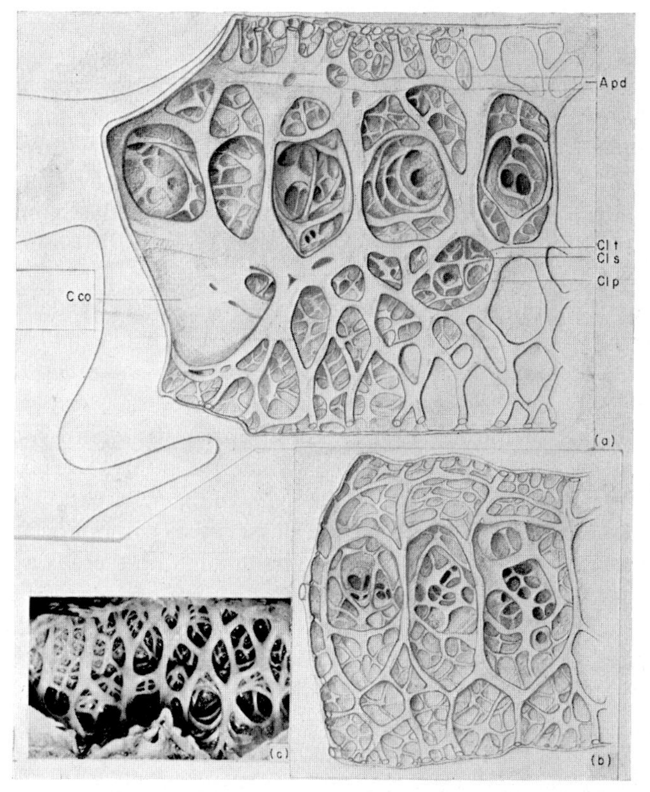

Fig. 4.7. Gross structure of the anterior region of the lung in *Protopterus aethiopicus* Heckel. (b and c): posterior region of the *Protopterus* lung showing extensive trabeculation. Apd: dorsal pulmonary artery; C co: communication cavity between the two lungs: Cl p: primary septum: Cl s: secondary septum, Cl t: tertiary septum (from Johansen, 1970).

A lungfish can, in contrast to most air-breathing fishes, use the lungs and the gills simultaneously in their respective respiratory media. The same condition prevails in many amphibians where pulmonary and cutaneous gas exchange occurs concurrently. This situation calls for circulatory adjustments to obtain optimal efficiency. If, for example, the P_{O_2} of pulmonary venous blood is far above that in ambient water, the O_2 picked up in the lungs might be lost to the water irrigating the gills or the skin surface. The functional anatomy of the circulatory system in lungfishes and amphibians has been described by Johansen and Hanson (1968), Johansen *et al.* (1968) and by Johansen (1970). An outline is given above (Fig. 4.6).

There are three species of lungfishes in existence today. The Australian *Neoceratodus forsteri*, the South American *Lepidosiren paradoxa* and the African *Protopterus aethipicus*. *Neoceratodus* is an obligate water breather, and has a single lung and gills that resemble those found in purely water-breathing teleosts.

The other two lungfishes are obligate air breathers with two lungs and vestigal gills with no secondary lamellae. A division of the heart has started, and the heart resembles more that of primitive tetrapods than of fishes (Bugge, 1961).

The importance of the lungs in *Lepidosiren* and *Protopterus* is shown by the fact that the arterial P_{O_2} is almost the same whether the animals stay in well-aerated water or in air. The arterial P_{CO_2}, however, is lower when the animal is in water (Johansen, 1970).

The state of gill reduction is reflected by the branchial vascular resistance. In *Neoceratodus* it is comparable to that of teleost gills, while in the two others it is very small and almost nil during air exposure (Johansen *et al.*, 1968).

The lungfishes have four pairs of branchial (gill) arteries originating from the anterior part of the heart. The ventral aorta is therefore almost non-existent. In *Neoceratodus* the four most anterior of these arteries supply normally developed gill arches with filaments and lamellae. In the other two the two anteriormost pass directly to the dorsal aorta without perfusing gills, the two posterior supply very reduced gills. In addition there is a fifth arch, a hemibranch, which is supplied by blood which has perfused the fourth gill arch. The efferent vessel from this hemibranch gives a branch to the pulmonary artery before it ends in the dorsal aorta.

A most important corollary development to the lung is a separate return of pulmonary blood, not to the general venous system, but directly to the heart. This is in contrast to the situation in other air-breathing fishes. Internal septa in the atrium and ventricle are anatomical evidence for partial separation of the two main venous sources, the deoxygenated vena cava blood and the oxygenated blood from the pulmonary vein (Fig. 4.6).

(2) *Pattern of blood flow.* The unique feature of the dual breathing in lungfishes is the ability to distribute oxygenated and deoxygenated blood to the appropriate vascular area depending upon the respiratory situation. We shall discuss the experimental evidence on this subject in some detail. Johansen *et al.* (1968) investigated the situation by measuring the P_{O_2} in input and output vessels of the heart. The P_{O_2} values were converted to O_2-content by the use of O_2-equilibrium curves (Fig. 4.8) and these values were used to calculate the degree of selective blood passage through the

heart, the gills, to the lung(s) and the systemic arterial system. Even though the anatomy of the fishes prohibits sampling from all localities of interest, the available measurements describe the situation quite well. Fig. 4.9 shows a schematic representation of the circulatory system, and indicates where blood samples were taken. The results of the blood gas measurements are shown in Table 4.1.

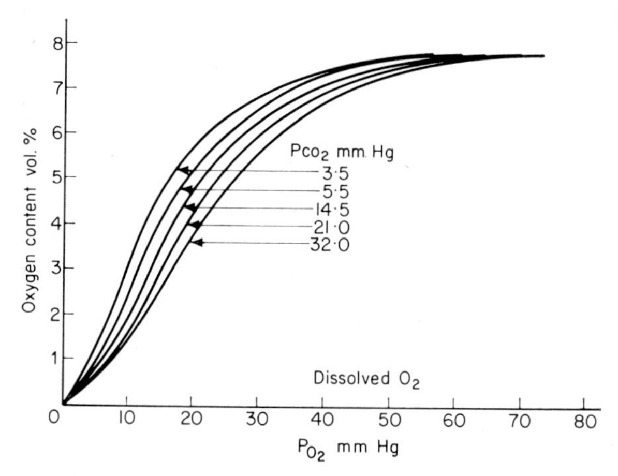

Fig. 4.8. Oxygen equilibrium curves of blood from the lungfish *Neoceratodus* at different P_{CO_2}. Temp: 18°C, haematocrit 36% (from Lenfant *et al.*, 1966/67).

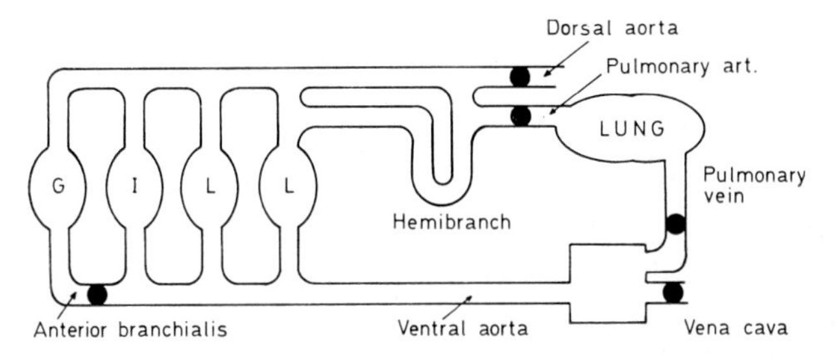

Fig. 4.9. Schematic representation of the circulatory system of lungfishes. Black dots indicate where blood samples were taken in experiments described in text.

When *Neoceratodus* stays in well-aerated water, it does not ventilate the lungs; consequently they can have no respiratory importance. This is substantiated by the observation that there is no pulmonary arteriovenous P_{O_2} difference and no gradient in P_{O_2} between pulmonary gas and blood. Pulmonary venous blood had a P_{O_2} of 38 mm Hg and vena cava blood 14 mm Hg. The anterior branchial artery had a P_{O_2} of 20 mm Hg. Converting these values to O_2-content (Fig. 4.8) we can calculate how much blood has come from the pulmonary vein and how much from the vena cava. Table 4.1 shows that they contribute approximately the same amount. This is somewhat

TABLE 4.1
Blood gas composition of two lungfishes. Oxygen content in volume per cent, P_{O_2} in mm Hg (from Johansen *et al.*, 1968).

				Pulmonary artery blood		Pulmonary venous blood		Anterior branch blood		Vena cava blood		Ratio of pulmonary venous blood to vena cava blood in:	
Species	No. of specimens	Condition	Systemic arterial blood	P_{O_2}-content	O_2-content	P_{O_2}-content	O_2-content	P_{O_2}-content	O_2-content	P_{O_2}-content	O_2-content	Anterior branch blood	Pulmonary artery blood
Neoceratodus	8	In aerated water	—	38	7·30	36	7·25	20	5·00	14	3·40	5/4	
	8	In hypoxic water	—	25	6·0	95	7·90	32	6·75	5	0·80	5/1	3/1
Protopterus	3	In aerated water	30–4·8	25	4·3	46	6·05	38	5·50	2	0·15	10/1	7/3

surprising since there does not appear to be any functional need for lung circulation under these conditions.

When *Neoceratodus* is transferred to hypoxic water (P_{O_2} = 40–80 mm Hg) it seeks the surface periodically and starts to ventilate the lung with air. Under these conditions the arterial pulmonary blood has a P_{O_2} of 25 mm Hg while the venous pulmonary blood has a P_{O_2} of 95 mm Hg, thus attesting to the importance of the lungs. Vena cava blood had a P_{O_2} of 5 mm Hg. Since the pulmonary arterial blood has passed the gills, it has been enriched in O_2 even at this low ambient P_{O_2}. The P_{O_2} of blood taken from the more anterior branchial arteries before they entered the gills was 32 mm Hg. This shows that there must be a selective distribution of more oxygenated blood to those arteries which supply the dorsal aorta, and less oxygenated blood to the pulmonary artery which supplies primarily the lung. We can now get a numerical answer to the question: in what proportion does the vena cava and the pulmonary vein contribute to the blood in the anterior branchial artery and in the pulmonary artery? The answer is (Table 4.1) that anterior branchial arterial blood is made up of about 5 parts pulmonary venous blood and 1 part vena cava blood. We can similarly calculate that the pulmonary artery blood is mixed from about 3 parts pulmonary venous blood and 1 part vena cava blood. Since pulmonary venous blood may have been oxygenated in the gills, it is likely that this relation in reality is closer to 2 : 1. There is thus a clear preference for pulmonary venous blood to enter the anterior branchial arteries and therefore the systemic arteries, whereas the pulmonary artery receives comparatively more vena cava blood.

Protopterus was similarly studied in aerated water. Under these conditions the anterior branchial arterial blood was composed of 10 parts pulmonary venous blood and 1 part vena cava blood while the pulmonary artery received about twice as much blood from the pulmonary vein as from the vena cava.

These measurements show that the lungfishes are able to pass oxygenated and deoxygenated blood selectively through the heart and dispatch it to different outflow routes. This selection acts to distribute oxygenated blood to the tissues and deoxygenated to that of the two respiratory organs which is functional. Another and rather puzzling result is that in all cases the blood flow through the lungs is larger than through the systemic vascular beds. This is particularly so when the animals rely exclusively on lung ventilation.

Functional separation of oxygenated pulmonary blood and deoxygenated vena cava blood in the heart and its outflow channels has been qualitatively established by Johansen and Lenfant (1967) for *Lepidosiren*. Thus the blood perfusing the dorsal aorta was almost saturated with a P_{O_2} of about 30 mm Hg while that in the pulmonary artery was about half saturated at a P_{O_2} of 15 mm Hg. Considerable mixing in the heart is evidenced, however, by the fact that the pulmonary venous blood showed a P_{O_2} of about 50 while the vena cava probably has a P_{O_2} of about 5 mm Hg (Johansen *et al.*, 1968).

The dual breathing pattern of *Protopteus* is paralleled also by changes in cardiac output. Johansen and Hanson (1968) showed that the pulmonary blood flow increases when the fish takes a breath of air. They also showed that in the phase just following a breath the separation of oxygenated and deoxygenated blood in the heart is maximal. This will on the one hand reduce the risk of losing O_2 via the gills, but will also cause variations in arterial P_{O_2} during the respiratory cycle.

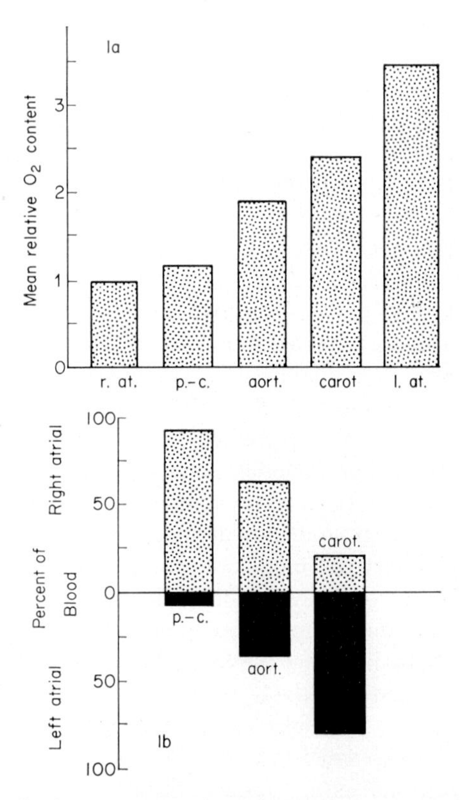

Fig. 4.10. Upper graph: Oxygen content in main vessels and heart chambers of the frog *Rana pipiens*. r. at. = right atrium, p.-c. = pulmo-cutaneous vessel, 1. at. = left atrium. Lower graph: The proportion of blood from right and left atrium in three main blood vessels based on data in upper graph (from DeLong, 1962).

Functional separation of oxygenated and deoxygenated blood is also found in amphibians. DeLong (1962) analyzed the O_2 content of blood from the main vessels entering and leaving the heart of the frog *Rana pipiens*. His results give conclusive evidence that the carotid artery contains predominantly well-oxygenated blood from the pulmonary vein, while the pulmo-cutaneous artery contains predominantly blood from the systemic veins.

However, the aorta contains mixed blood (Fig. 4.10). A similar situation has been described for other amphibians (Johansen and Hanson, 1968).

(3) *Lungfishes, dual respiration and the four-chambered heart.* It is noteworthy that while the principal solution to bimodal respiration and the partially divided heart of lungfishes represents the evolutionary direction leading to vertebrate air breathers, the "electrophorus-solution" represents an evolutionary dead end. One might ask: why was not the latter line further explored? The gills could be further reduced and the mouth papillae further developed with blood drainage into the dorsal aorta. This could in time have led to an air breather with a two-chambered heart. Since such animals do not exist something must be fundamentally unsatisfactory with this solution. The following discussion may bring forth some pertinent points.

The circulatory system of homeotherms is distinguished by at least one outstanding feature: the simultaneous operation of a high-pressure systemic circulation and a low-pressure pulmonary circuit. This is made possible by the complete division of the heart in two separate, although frequency synchronized, parts. The high-pressure circuit enables rapid gas transport over large distances, thus allowing animals to grow big. Simultaneously, the low-pressure circuit allows the blood to be separated from the air by a thin and mechanically fragile barrier without extensive oedema formation. This aspect may have become important very early. It is noteworthy, that even in the least air breathing of lungfishes, *Neoceratodus*, the blood pressure is considerably lower in the pulmonary artery than in the ventral aorta (Johansen *et al.*, 1968). This seems chiefly to be due to the considerable resistance caused by the branchial circulation. However, also in the two other lungfishes, which have well-developed lungs and thoroughfare vessels for shunting blood around exchange vessels in the gills, the pulmonary arterial pressure is lower than the blood pressure in the ventral aorta. Also, the pulmonary vascular resistance is less than the systemic. Thus, at this early stage the development of lungs is accompanied by reduced pulmonary blood pressure. On the other hand, in air-breathing teleosts both gills and accessory organs are supplied directly from the heart and the systemic arterial pressure is equal to the pulmonary.

It may appear, therefore, that the first step towards the four-chambered heart was to solve the gas exchange problem, not by modifying the gills or the mouth cavity, but by building a lung at a certain distance from the gills. This solution implied the possibility of using both organs simultaneously, but utilizing different respiratory media. It also rewarded developments towards functional separation of pulmonary venous blood from systemic venous blood. When further developments towards pure air breathers were "tried", based on the "electrophorus" and the lungfish types, it turned out that the partial separation of the heart, which in lungfishes had been selected for respiratory purposes, was the key for the acquisition of the low-pressure circulation of lungs. It is interesting that large reptiles and long snakes, both of which need high systemic blood pressure and have some transventricular communication, maintain, by muscular contraction, a gas pressure in the lungs higher than the ambient pressure. This might be important to maintain a low hydrostatic pressure gradient across the pulmonary exchange membrane.

Only three genera of lungfishes are in existence today. Apparently they had their mission as transitional forms, not as end forms. Many more species of air-breathing fishes with the mouth or gut full of respiratory tissue survived. As organisms they may thus be more successful than lungfishes, but they represent a dead-end line in evolution.

(4) *Ventilation-perfusion relations in dual breathers.* To function optimally there must exist a strict relationship between ventilation and perfusion of a respiratory organ. This is a well-known co-ordination in mammalian lungs. In elasmobranchs Satchell (1960) found evidence for controlled coupling between ventilation and circulation in the gills and pointed out the significance of this interplay for an optimal functioning of, for example, countercurrent gas exchange. The dual respiratory function of *Protopterus* also

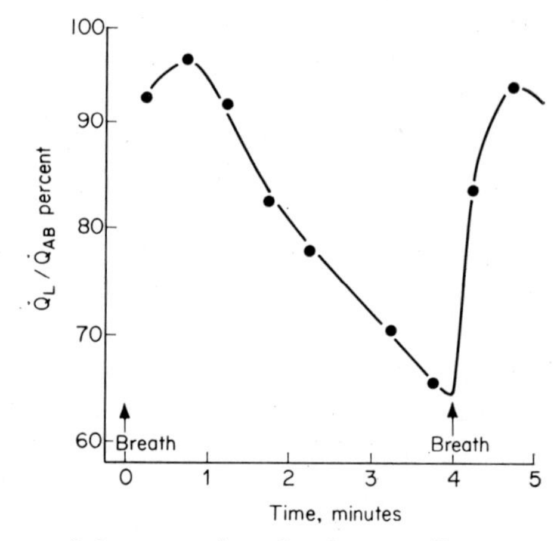

Fig. 4.11. Time course of the proportion of pulmonary flow to total flow perfusing the anterior gill-less branchial arteries during an interval between air breaths in *Protopterus aethiopicus* (from Johansen *et al.*, 1968).

illustrates this point. Fig. 4.11 shows that the proportion of the total blood flow perfusing the lungs is far higher just after a breath than later in the cycle. This is in accord with the finding (Johansen and Lenfant, 1968) that the P_{O_2} gradient between lung gas and blood is highest just after a breath. The circulation time at that point appears to be too short to allow complete saturation. This has the consequence that the arterial P_{O_2} declines less during the apnoeic (breath-holding) period than does the lung P_{O_2}. This assists in the maintenance of a constant arterial P_{O_2}.

(5) *Blood properties in dual breathers*. The transition from aquatic to aerial respiration is necessarily reflected in the P_{CO_2} of circulating blood. True air breathers exchange CO_2 only through the lungs where, owing to two-way ventilation, the P_{CO_2} is as much higher than the atmospheric P_{CO_2} as the P_{O_2} is below it. Thus typical arterial blood P_{CO_2} in air breathers is 40–50 mm Hg. Typical water breathers exchange both gases with water. Because CO_2 is about 25 times as soluble as O_2 in water, the typical P_{CO_2} in their blood is below 5 mm Hg while the P_{O_2} may still be 50–60 mm Hg. Transitional forms that exchange gases partly with lung-gas and partly with water, through gills or skin, will necessarily have intermediate P_{CO_2} values. This is often measured by the exchange ratio (volume of CO_2-exchanged to volume of O_2-exchanged) of the respiratory organ. Animals exchanging with one organ only show exchange ratio equal to the RQ. Dual breathers show high exchange ratios for the aquatic organ (Johansen, 1970).

Both *Symbranchus* and the obligate water-breathing lungfish *Neoceratodus* have arterial P_{CO_2} below 10 mm Hg when in water, but somewhat higher when in air (Lenfant *et al.*, 1966/7). The bladder-breathing *Amia* shows a similar pattern, although the arterial P_{CO_2} is lower (Johansen *et al.*, 1970). The electric eel which is primarily an air breather, has P_{CO_2} above 15 mm Hg while the typical air-breathing lungfishes *Protopterus* and *Lepidosiren* show arterial P_{CO_2} above 20 mm Hg (Lenfant and Johansen, 1968; Johansen and Lenfant, 1967).

In Chapter 3 it was pointed out that fishes which live in hypoxic and often acid water typically have blood with higher O_2-affinity and a smaller Bohr-shift than fishes from well-oxygenated water. Dual breathers are typically found in hypoxic and acid water. The P_{50} of blood from these animals ranges from 15 to 20 mm Hg at 20°C and normal pH and P_{CO_2}. This is higher than that normally found in pure water-breathing fishes from similar habitats.

This relatively high P_{50} of dual breathers seems reasonable since these fishes are not prisoners of their aquatic environment but may switch to aerial gas exchange when the conditions call for it.

The Root-shift of fishes, whether dual breathers or water breathers, from these habitats is either small or altogether absent (Ch. 3,3.B). Dual breathers normally show slight Bohr-shifts, the O_2-equilibrium curves of *Neoceratodus* (Fig. 4.8) are typical.

Since the transition to aerial breathing occurred in O_2-deficient water, it is to be expected that the O_2-affinity of the blood from these transitional forms should be lower, i.e. the P_{50} higher, the more developed the air-breathing mechanisms are. This would be in accord with the parallel that is present between P_{50} and environmental P_{O_2} in "normal" fishes. However, the three species of lungfishes have very similar P_{50}, the most aquatic, *Neoceratodus*, having in fact the highest P_{50} (Fig. 4.12). In a series of amphibians, the most aquatic has a P_{50} about one-half of that in the totally aerial species (Lenfant and Johansen, 1967) (Fig. 4.13).

When we compare the O_2-affinity of haemoglobin of aquabreathers from well-oxygenated water (salmon, mackerel, etc.) with that from pure air-breathing poikilotherms, no clear difference is discernible. Thus the nature of the milieu does not by itself seem to be reflected in the O_2-affinity. This

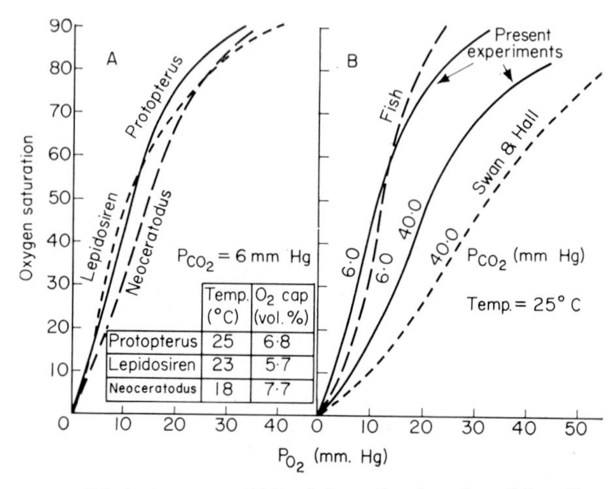

Fig. 4.12. Oxygen equilibrium curves of blood from the three lungfishes (from Lenfant and Johansen, 1968).

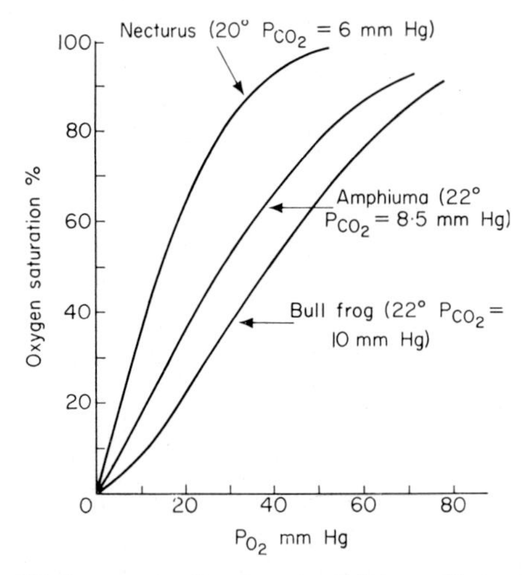

Fig. 4.13. Oxygen equilibrium curves for three amphibians. *Necturus* is the most aquatic while the bullfrog relies predominantly on aerial respiration (from Lenfant and Johansen, 1967).

indicates that gas exchange across gills and lungs occurs at about the same gradient, which is a reassuring result in view of the fact that the exchange barriers of the two types of organs are of the same dimensions.

Neither does there seem to be any clear correlation between the type of environment and the blood O_2-capacity of the inhabitants. Reptiles, amphibia and fishes all show O_2-capacities of about 10–15 volume per cent at physiological pH-values. It may be interesting that the O_2-capacity is distinctly lower in lungfishes than in air-breathing fishes from the same habitat. The high O_2-capacity in the blood of *Symbranchus*, *Anguilla* and *Electrophorus* compared with that found in lungfishes, may be a compensation for the less efficient respiratory mechanism implicit in the poorer separation of oxygenated and deoxygenated blood (Lenfant and Johansen, 1968).

B. Respiratory Adaptations Accompanying the Amphibian Metamorphosis

The transition of amphibia from aquatic eggs via tadpoles to air-breathing adults is accompanied by changes in the respiratory processes. The amphibian egg does not exhibit structural specializations for gas exchange and therefore obtains ample gas exchange through the general surface. The tadpole shows

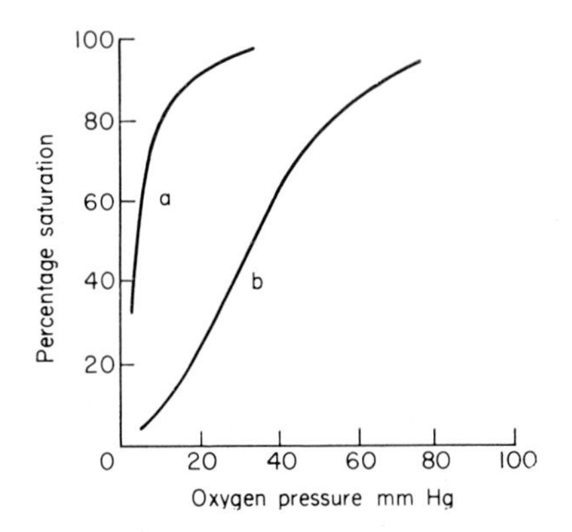

Fig. 4.14. Diagram to show the form of: (curve a) the rectangular hyperbola and (b) the sigmoid curve typical of the oxygen dissociation curves of tadpoles and of adult frogs respectively (from Foxon, 1964).

first external gills, then internal gills and, later in the metamorphosis to adult frog, also lungs. The development of these structures varies among amphibia, but typically follows the course described above.

The degree of vascularization of skin, gills and lungs during the developmental stages indicates the relative importance of these areas in gas exchange. Strawinski (1956) investigated this problem in the frog *Rana esculenta*. The density of capillaries in the skin increases during development of the tadpole and reaches a maximum at the stage when the forelimbs break through. This density is maintained in the adult skin although the distance from blood to the exterior is increased.

As has been mentioned earlier in this chapter the adult amphibian heart exhibits selective passages of oxygenated and deoxygenated blood. Blood passage through the heart at other developmental stages has not been investigated.

Drastich (1927) showed that in salamander larvae the development of the gills depends on the O_2-concentration of the environment. The more O_2-deficient the more developed the gills. Bond (1960) found that in O_2-poor water the gill would not only be larger but also better adapted for gas exchange. Thus with enlargement of the gills there is also flattening of the epithelial cells.

Metamorphosis usually begins in rather O_2-deficient water and proceeds in more O_2-rich habitats. Even though the eggs are laid in the same pond where the tadpole develops, the P_{O_2} in the egg cluster or in a mass of tadpoles may be considerably decreased. Thus Savage (1935) found the water 50 cm outside the egg cluster of *Rana temporaria* to be 136% saturated, whereas inside the spawn it varied between 3 and 16%. Tadpoles of this species tend to aggregate. When the tadpole transforms into the adult form and develops lungs it has access to the far more O_2-rich air. This change of respiratory habitat is paralleled by changes in the blood O_2-affinity. McCutcheon (1936) found a P_{50} of 5 mm Hg for the tadpole of the bullfrog *Rana catesbeiana*. At this P_{O_2} the blood of the adult frog was only 5% saturated (Fig. 4.14).

Chapter 5

AERIAL GAS EXCHANGE

We can distinguish three typical mechanisms for aerial respiration but intermediate forms are frequently found:

(A) gas exchange across the general body surface with circulation and respiratory pigment
(B) gas exchange by tubes which lead air directly to the tissue
(C) gas exchange with lungs and a circulatory system with respiratory pigments.

In addition we find, as was the case for aquatic animals, many small organisms where no distinct respiratory or circulatory system can be recognized. This situation prevails in many parasites and small terrestrial arthropods.

A. Gas Exchange Across the General Body Surface with Circulation and a Respiratory Pigment

This type of respiratory mechanism is exemplified by many annelids. They have no distinct respiratory organ, and gas exchange occurs across the body surface (Fig. 5.1). This surface is, however, of varying thickness and gas exchange probably occurs primarily across certain areas. Annelids have a closed circulatory system and haemoglobin of molecular weight from 2 to 3×10^6 dissolved in the plasma. In some annelids, the earthworm is an example, there is also intracellular Hb.

The respiratory mechanism of several species of earthworms have been studied (Laverack, 1963). Their blood is characterized by a high O_2-affinity (P_{50} from 2–8 mm Hg). The giant earthworm (*Glossoscolex giganteus*) which may weigh up to 600 g and have a diameter of 2–3 cm and a length of 120 cm, has a P_{50} at 20°C and pH = 7·50 of 7 mm Hg (Johansen and Martin, 1966). The O_2-equilibrium curves have a conspicuously sigmoid shape but show only a slight Bohr-shift (Fig. 5.2). The O_2-capacity is high (8–12 volume per cent). Haughton *et al.* (1958) investigated the blood of two other species of earthworms and found similar results. The O_2-affinity also showed a high temperature sensitivity (Fig. 5.3).

The P_{O_2} in soil similar to that inhabited by earthworms varies between 110 and 150 mm Hg (Hack, 1956). However, considerably lower values probably exist.

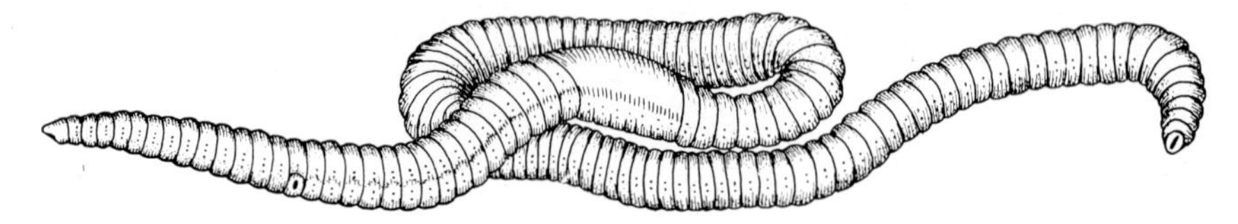

Fig. 5.1. The respiratory organ of an earthworm.

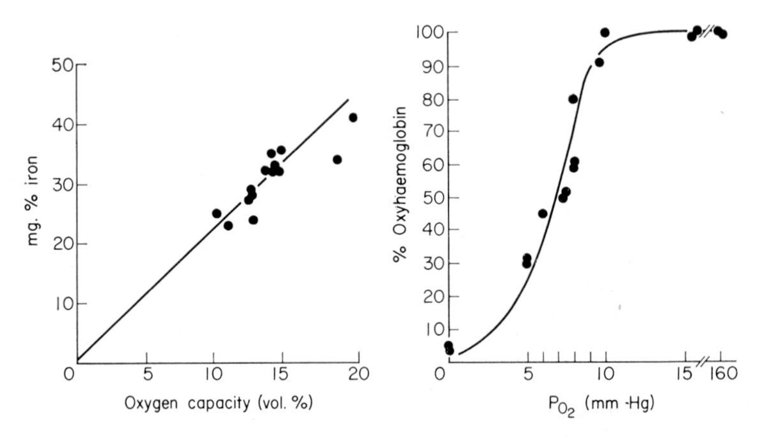

Fig. 5.2. The O_2-capacity (left) and O_2-equilibrium curve of blood from the giant earthworm, *Glossoscolex giganteus*. Temperature 20°C, pH = 7·8 (from Johansen and Martin, 1966).

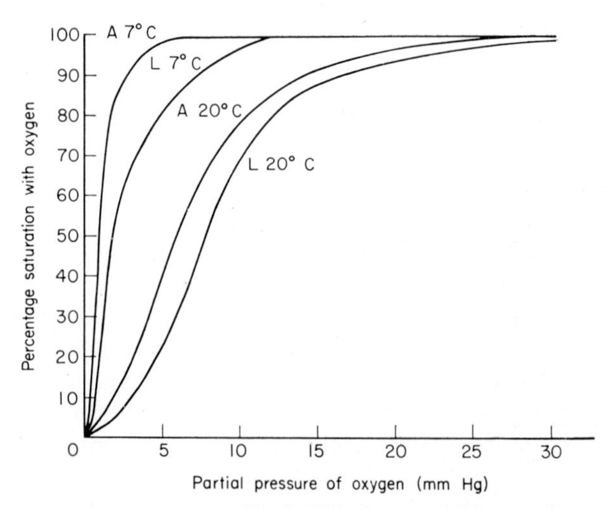

Fig. 5.3. Oxygen-equilibrium curves of haemoglobin from *Lumbricus* and *Allolobophora*. The curves for *Allolobophora* lie to the left of those for *Lumbricus* at the corresponding temperatures. *A* refers to *Allolobophora*, *L* to *Lumbricus* (from Haughton *et al.*, 1958).

The transport function of the Hb in these animals has been questioned since the P_{50} is so low. However, measurements of O_2-uptake in the presence and absence of CO, by Johnson (1942) and later by Cosgrove and Schwartz (1965), have definitely established that Hb is an important carrier of O_2 under *in vivo* conditions. The same investigators also concluded that about 30% of the total O_2-consumption was transported by Hb (Fig. 5.4). This is surprisingly low in view of the high blood O_2-capacity. It thus appears that as the blood circulates the tissues only a small proportion of its O_2-content is removed. However, since the P_{50} is low this should not change the P_{O_2} much.

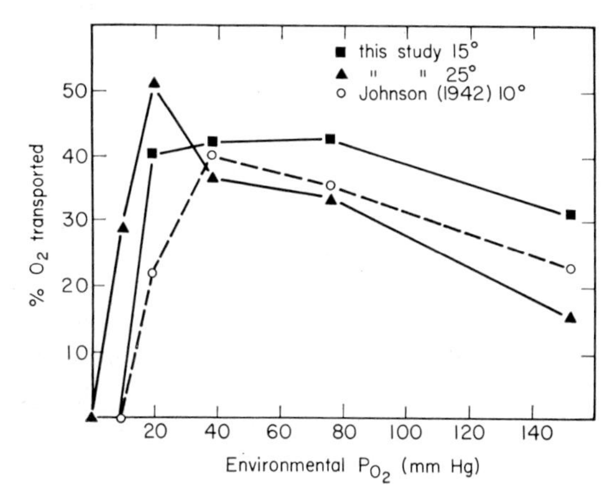

Fig. 5.4. Percentage of total O_2 consumption that is transported by the respiratory pigment in earthworms at different environmental P_{O_2} (from Cosgrove and Schwartz, 1965).

Such a mechanism with high O_2-capacity, low P_{50} and low O_2-extraction may well be an adaptation to the respiratory situation of earthworms. Since their respiratory surface is part of the body surface, body movements will most likely influence the degree of exposure and therefore of O_2-exchange. The O_2-supply will therefore vary and the high O_2-capacity blood may serve as an O_2-reservoir much like myoglobin of higher animals.

Earthworms also have a rather unique circulatory system. Blood from the exchange areas is mixed with deoxygenated blood before being presented to the tissues, and considerable variation in the degree of mixing appears probable (Johansen and Martin, 1966).

A respiratory system characterized by high O_2-capacity, high O_2-affinity and a large circulatory flexibility thus appears to be a reasonable arrangement for an animal that experiences a highly variable opportunity of O_2-exchange.

The presence of a pigment in solution will increase the viscosity and the colloidal osmotic pressure of the blood. It may be significant that blood from both annelids and cephalopods has dissolved Hb with a molecular weight from 2 to 9×10^6 and an O_2-capacity from 4 to 12 volume per cent, while crustacean blood contains dissolved pigment of molecular weight below 10^6 and O_2-capacity below 1·5 volume per cent.

Even in animals with specialized respiratory organs the general body surface will more or less take part in gas exchange. This is particularly important in animals with moist skin but will, to a varying extent, take place in all animals. The fact that, in man, gas exchange through the skin amounts to only 0·2% of the total (Krogh, 1941), has little respiratory significance. This low gas exchange is due, in part, to a very O_2-impermeable layer in the epidermis (Sejrsen, 1968).

Gas exchange through the skin of fishes during air exposure has already been discussed (Ch. 4). Chaetum (1934) observed that naked gastropods of the genera *Lymnea* and *Helicostoma* can stay submerged in water for long periods of time while their lungs are filled, but not ventilated, with water. Thus gas exchange with water through the skin seems at least temporarily to suffice. The same is most likely the case when the animals are exposed to air although experimental evidence is lacking as to the relative importance of skin and lung respiration in air.

The relative importance of skin and lung respiration during air exposure of some lower vertebrates is well illustrated in the studies by Krogh (1904) and by Dolk and Postma (1927) on frogs and by Whitford and Hutchinson (1965) on salamanders. Krogh found that despite large variations in total gas exchange (70–170 ml/kg/h) the oxygen uptake through the skin remained practically constant at about 50; the variations were therefore brought about by changes in the ventilation and the blood flow of the lungs. Carbon dioxide which diffuses more rapidly than O_2 was eliminated at all metabolic levels mainly through the skin, where blood is exposed to an atmosphere practically CO_2-free. CO_2-elimination through the lungs is limited by the alveolar CO_2-tension and is therefore closely correlated with the ventilation. This situation is identical to that in lung fishes and indeed in all lung-breathing aquatic animals without an outer impermeable cover like cuticle, hair or feathers.

The fact that some salamanders (e.g. *Desmognathus quadramaculatus*) exist without lungs or other respiratory organs shows that cutaneous gas exchange may indeed be sufficient. Czopek (1966) reviews a series of investigations of the functional anatomy of capillaries in the skin, the lungs and the mouth of amphibians. Species with poorly-developed lungs had thinner epidermis and a denser capillary network in the skin than species with well-developed lungs.

B. Gas Exchange by Tubes Which Lead Air Directly to the Tissues

1. THE ARCHITECTURE OF THE TRACHEAL SYSTEM

This subject has been reviewed by Miller (1964). A thorough anatomical and functional analysis of the tracheal system of locusts and dragon-flies has been described in three papers by Weis-Fogh (1964 a, b and c).

Although the anatomy of the tracheal system varies greatly from species to species, the basic lay-out is uniform. The tracheal system consists of a network of branching, air-filled tubes ending blindly in the tissues. The tracheal air communicates with the atmosphere through pairs of spiracles. The structure of the spiracles varies greatly. In primitive insects (e.g. *Peripatus*) it is a simple hole whose aperture cannot be regulated. In most insects, however, it is more complex and can be opened or closed according to physiological demand. In some insects the spiracular opening is covered by a perforated plate situated inside the spiracular valve.

This valve guards the entrance to the primary trachea, which gives off branches at right angles forming secondary tracheae. These run parallel to muscle fibres and branch into tertiary tracheae which again subdivide to form the tracheoles. The two latter types of air passages thus are in very intimate contact with the tissues and the tracheoles may even indent cells like a finger in a rubber balloon.

The tracheoles are traditionally considered the main site of gas exchange, although there is no evidence of higher gas permeability of tracheoles than of trachea. Their wall, the intima, is very thin (10–50 mμ) while that of the trachea is often about twice as thick; the tracheolar intima is often smooth as opposed to that of the trachea which is papillate; in tracheoles the coiled ridges seem to form single helices while in tracheae double helices are common. Furthermore the tracheolar intima is not shed, as is the tracheal one in late moults. However, the great number of tracheoles and thereby enormous total wall area and close association with cells, justify our considering them analogous to vertebrate capillaries as the main sites of gas exchange.

The tracheal lining appears more hydrophobic than the tracheolar one. This has been taken as showing that tracheoles are more gas permeable than tracheae. It has been shown, however, that lipid films need not restrict CO_2-diffusion significantly (Blank and Roughton, 1960). Most important, however, is the fact that the lipid lining covering the intima is so thin compared to the rest of the tissue separating gas from sites of metabolism, that even with a low permeability it would hardly contribute significantly to the overall diffusion resistance.

The tracheal system occupies from 5 to 50% of the volume of insects. The proportion varies among species and according to the age and developmental stage of the individual. There seems to exist a positive correlation between the tracheation of an organ and its supposed metabolic activity, in much the same way as between capillarization and metabolic activity in vertebrates.

We shall notice the structural similarity between the tracheal system of insects and the air system of bird lungs. A comparison of the functional design of the two systems is described by Weis-Fogh (1964a).

2. GAS RENEWAL OF THE TRACHEAL SYSTEM

The entire tracheolar system is not always air filled. In his classic experiment on mosquito larvae Wigglesworth (1930) showed that during rest the tips of the tracheoles were filled with fluid, while during activity they were empty (Fig. 5.5). The mechanism of this change is still debated. Wigglesworth's original explanation was simple and direct: during high metabolism metabolites will accumulate and increase the osmotic pressure of the fluid surrounding the tracheoles. Water will therefore diffuse out of them and leave the tracheole empty. During rest the surrounding osmotic pressure decreases and water should diffuse back in again. There are some defects in this story,

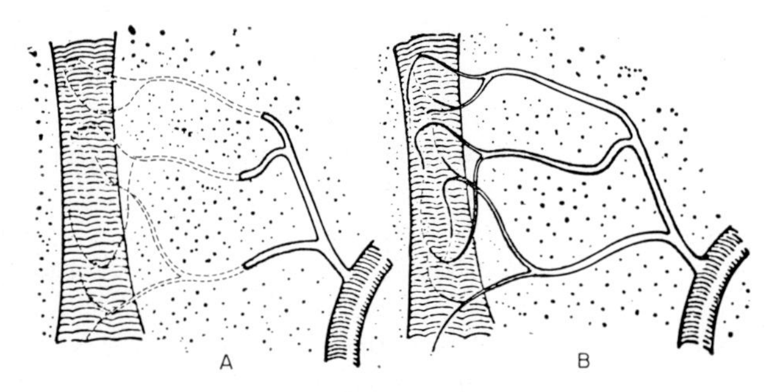

Fig. 5.5. Tracheoles running to a muscle fibre; semischematic A, muscle at rest; terminal parts of tracheoles contain fluid. B, muscle fatigued; air extends far into tracheoles (from Wigglesworth, 1930).

however. The main one concerns the osmolarity of the fluid in the trachea. If it is pure water it would always diffuse into the tissue, if it is a saline solution partial water diffusion into the tissue would increase its osmolarity and stop further osmosis. To meet this difficulty several mechanisms have been suggested (Wigglesworth, 1953) of which active ion transport, differences in ion and water permeability and different permeabilities in the two directions of the target cells seem most promising.

In some species the tracheal system is not ventilated, and gas renewal appears to occur by diffusion only. Krogh (1920) calculated that the dimensions of the tracheal system of the *Cossus* larvae would allow sufficient O_2-supply to the tips of the tracheoles by diffusion only (Fig. 5.6). Buck (1962) re-examined this problem with special attention to the diffusion path from the tracheolar gas to the sites of O_2-consumption. Owing to the 10^6 times higher diffusion rate in gas than in water, this latter barrier may in fact be rate limiting. By comparing the architecture of the tracheolar system, as revealed histologically, with calculations on limiting diffusion distances, Buck concludes that in some cases gas exchange can occur by diffusion alone while in others, and these are probably more frequent than previously assumed, a certain degree of ventilation is necessary. Owing to the uncertainties involved in a quantitative treatment of this problem, it may be most realistic to accept that ventilation signals a functional need even if our calculations indicate that it is unnecessary.

In many species the tracheal system is ventilated. The pattern of ventilation varies from species to species and within one specimen depending upon the conditions. In some species the same spiracle is used for both inspiration and expiration; however, in most cases inspiration occurs through the anterior

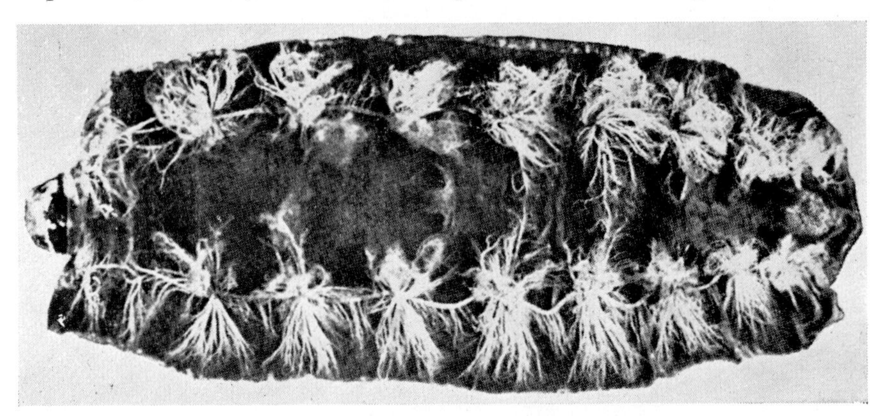

Fig. 5.6. A photograph of the tracheal system of the *Cossus* larvae (from Krogh, 1941).

spiracles and expiration through the posterior ones. In grasshoppers and locusts (Miller, 1960; Wigglesworth, 1953) at rest, inspiration occurs via the four anterior pairs of spiracles, while expiration occurs via the six posterior pairs. The flow is driven by alternating dorsoventral compression by two sets of muscles.

Ventilation, whether tidal or unidirectional, is often very regular, but sometimes it appears to be completely irregular. Many immature individuals show ventilatory pauses, and in working adults supplementary ventilations occur. During flight ventilation may be aided by direct inflow of ambient air at the ventral surface and a slight underpressure over the abdominal spiracles owing to a Bernoulli-Venturi-effect (Stride, 1958).

Among cockroaches, *Periplaneta* and *Blatella* make no ventilatory movements during rest, but ventilate tidally under stress. Other roaches *Byrsotria*, *Blaberus* and *Nyctobra* show anterior-posterior ventilation also during rest (Buck, 1962).

3. GAS EXCHANGE DURING FLIGHT

This subject demands special attention for at least three reasons. One is that the O_2-consumption during flight is 10–100-fold higher than during rest (Weis-Fogh, 1961). Understanding of the respiratory mechanism that permits this might therefore suggest general properties of especially adaptive systems. Another reason is that the absolute rate of O_2-consumption of insect flight muscles during flapping flight is the highest recorded in any tissue (7·3 ml O_2 per g tissue per min). A third reason is that the problem is extraordinarily well analysed by Weis-Fogh (1964, a, b and c). The subject has been reviewed by Miller (1966).

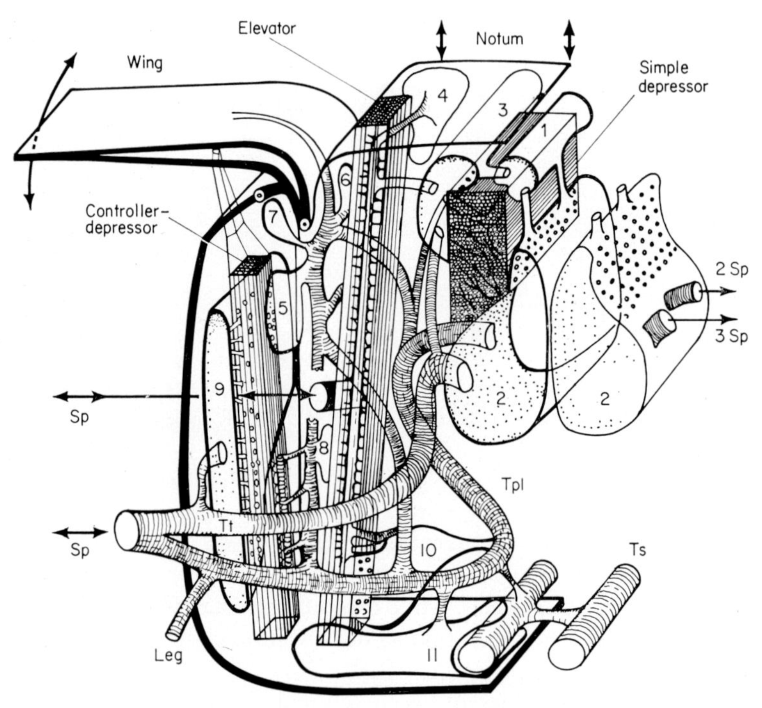

Fig. 5.7. Illustrates the architecture of the tracheal system. This is a narrow transverse section of the right side of the pterothorax of *Schistocerca gregaria*, opened laterally, to show the tracheal supply to the flight muscles. Arrows indicate the movements of the wings and notum. Three muscles are shown, a depressor, an elevator and a controller depressor. The tergo-pleural (Tt) and sternal (Ts) tracheae are shown with transverse hatching, while the pleuro-coxal tracheae (Tpl) are additionally marked with longitudinal broken lines. Air sacs 1 and 2 belong to the mesial group of the tergo-pleural system, and 3, 4, 5, 6 and 7 to the intermediate group. Sac 9 represents the lateral subdermal group, and sac 11 the ventral subdermal system. Sac 10 represents the coxal group of the pleuro-coxal system. The only anastomosis between sternal and pleuro-coxal systems shown is blocked by liquid (from Weis-Fogh, 1964a).

Weis-Fogh studied gas exchange of the flight muscles in the locust (*Schistocera gregaria*) and in dragon-flies (*Aeshnia*, *Libellula* and *Sympetrum*). The tracheal system of the left and right halves of the thorax is almost completely separated. The tracheal system originating from one spiracle communicates only with other spiracles on the same side (ipsilateral spiracles). Fig. 5.7

indicates the pattern of tracheae while Fig. 5.8 is a schematic drawing. During rest the system is unidirectionally ventilated by abdominal pumping, while ventilation during flight is achieved mainly by dorsoventral movements of the nota (the hard dorsal plates of the thorax). These movements cause ventilation of the primary and secondary trachea and of the air sacs. The tertiary trachea and the tracheoles are not ventilated. Instead gas exchange between the ventilated tracheae and the tissues occurs solely by diffusion. Weis-Fogh substantiated this assumption by calculations based on various models. Such a procedure is particularly meaningful in these insects since the tertiary trachea and tracheoles are arranged in a regular pattern.

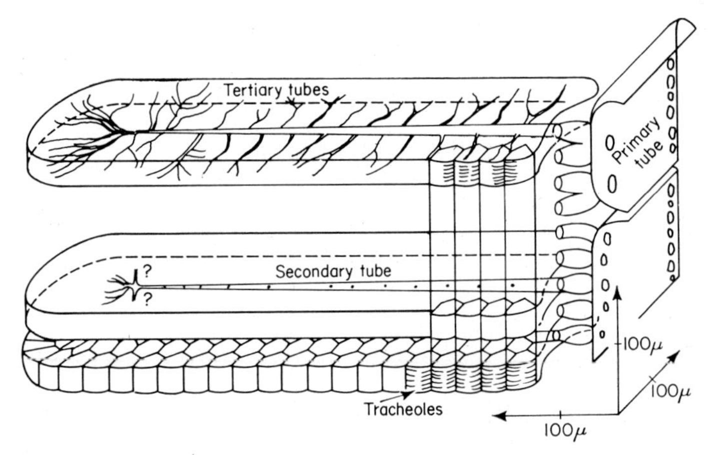

Fig. 5.8. Schematic reconstruction of the tracheal supply to a lobe of a wing muscle in a dragon-fly (from Weis-Fogh, 1964b).

The calculations were based on observed values for the tracheal system and its relation to the tissue cells. He concluded:

1. The O_2 present in the primary tracheae will suffice for only a few muscle contractions, therefore an efficient renewal is necessary. In small insects like *Drosophila* diffusion will suffice even during flight, but in larger species like locusts and dragonflies ventilation is required.

2. Beyond the primary tracheae and air sacs diffusion in the airways will suffice in all but the largest insects. In these, ventilation of secondary trachea has been observed.

3. The efficiency of diffusion in the tracheal system is due to the large *hole fraction* (ratio of the cross-sectional area of trachea to total area of a cross-section of the tissue) which in wing muscles varies between 10^{-1} and 10^{-3}.

4. At a reasonable ΔP_{O_2} of 0·05 atm between air in the primary tracheae and in the tips of the tracheoles, the theoretical maximum distance compatible with diffusion as the only mechanism of O_2-removal is calculated at

different levels of O_2-consumption. These distances are longer than those directly observed.

At constant O_2-consumption of 4 ml O_2/g/min the limiting tube length is higher, as is the hole fraction. In the locust with wing muscles having a hole fraction of 0·07 the theoretical tube length limit at these conditions is above 1 mm. At a hole fraction of 10^{-3} the limit is about 0·1 mm. The exact numbers depend upon which model the calculation is based.

5. The main part of gas exchange between gas and tissue occurs across the tracheoles. The distances between tracheoles and sites of metabolism allow ample gas exchange to occur by diffusion.

At a maximum rate of O_2-consumption of 8 ml O_2/g/min the maximum distance which is compatible with diffusion is 3·6–5·9 μ, depending on the arrangement of tracheoles. Direct observations indicate that these distances are not exceeded in tissues known to have O_2-consumption of this magnitude.

4. COMPARISON BETWEEN THE TRACHEAL SYSTEM AND THE VERTEBRATE CAPILLARY SYSTEM

Since the tracheoles of insects are analogous to capillaries of vertebrates, it seems appropriate to compare some aspects of the two systems. As an

TABLE 5.1

Quantitative information on a dog muscle (from Krogh, 1922).

Weight kg	Capillaries per mm²	Average distance from centre of capillary to halfway to the next capillary μ	Average diameter of capillaries μ	cm² capillary surface per cm³ muscle	fraction of cross-section occupied by capillaries (Hole fraction)
5	2600	11	7·2	590	10^{-1}

example of a quantitative treatment of capillaries I chose the classical one of Krogh (1922). In a dog muscle he found the data shown in Table 5.1. According to Krogh's calculations based on these dimensions, a ΔP_{O_2} of only 2–3 mm Hg between capillary blood and tissue should suffice for an adequate O_2 supply. Krogh concludes that during such hard muscular work circulation takes place through a much larger number of capillaries than appears necessary to ensure the supply of O_2. However, the fact that venous blood from working muscles may be practically devoid of O_2 throws doubt on this conclusion. If the same blood flow should perfuse half the number of capillaries, the O_2-extraction from the blood could not possibly be equally high.

It is also interesting to notice that the diffusion distances for O_2 in insect muscle and dog muscle are about the same, but the tracheal system can supply about 10 times more O_2 per g tissue than can the blood capillary system. This must be due to a difference in gradient, which again indicates that O_2-supply by diffusion through air tubes is more efficient than transport by the blood stream. The efficiency of the latter depends, at maximal capillarization, upon the rate of blood flow, which therefore appears to be the limiting parameter.

5. GAS EXCHANGE AND WATER ECONOMY—CYCLIC CO_2-RELEASE

A number of insects live in a dry atmosphere with limited or no access to water. Desiccation is thus an ever imminent companion of gas exchange. The following discussion will show how insects developed an optimal compromise between gas exchange and water economy. We should be aware that no insect, nor any other animal, has solved the problem so that it can combine a high degree of gas exchange with no water loss.

Two principles of water conservation have been used by insects. One is to develop body surfaces with a low water permeability. The other is to shield the respiratory surfaces from direct exposure to the atmosphere and thereby obtain a situation where water loss is limited, not by the rate of evaporation, but by diffusion between the gas and the atmosphere. This is accomplished by the narrow spiracular openings.

As to the first point the cutaneous gas exchange in adult air-exposed insects rarely exceeds 10% of the total O_2-uptake. The cuticle is often covered by a coating which appears to be far less permeable to water than to O_2. Thus Ito (1953) and Richards (1957) have demonstrated that scratching the cuticle and thereby partly removing the coating, increases cuticular water loss much more than cuticular gas exchange. Tuft (1950) reported that the wax coating of the *Rhodnius* egg is very O_2-permeable. It appears, therefore, that the generalization that permeability to water and to O_2 always parallel each other does not apply to such coatings.

The other mechanism which entails reduced water loss is a corollary to what is often termed "cyclic CO_2-release". Heller (1930) was the first to observe that CO_2 is intermittently released from butterfly pupae. Through the development of more and more refined methods such cyclic CO_2-release has been observed in many species (Buck, 1962) (Fig. 5.9). The frequency varies a lot, from once every other minute to one burst every 25 h. The higher the frequency, the shorter the duration of a burst. Insects exhibiting such behaviour all have occlusible spiracles and appear to have a high P_{CO_2} spiracular triggering level. It is also important to notice that all these insects have a rather low rate of metabolism. Thus bursts are in many species observed during rest, but absent during activity or other conditions of high metabolism (Buck, 1962). Thus the burst mechanism is not in operation during exercise.

As a rule a burst is signalled by opening of the spiracles. Normally no ventilatory movements are observed during a burst. Gas exchange therefore occurs primarily by diffusion through the open spiracles. During the inter-burst period gas exchange occurs by diffusion across the closed spiracles. Kanwisher (1966) investigated gas and water exchange in the *Cecropia* pupae in an admirably simple and direct manner. He made continuous records of CO_2 and water release from the pupae as well as O_2-uptake. In addition he measured tracheal volumes and the composition of tracheal gas.

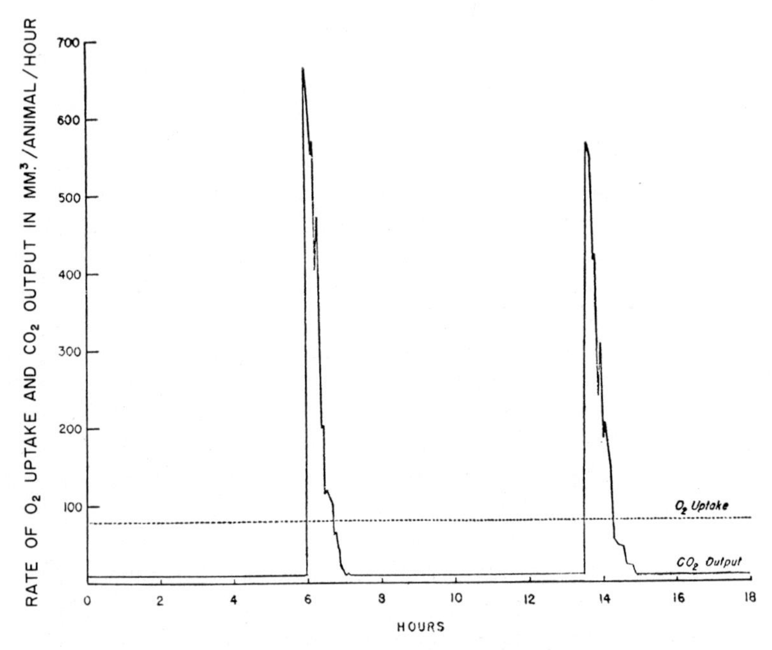

Fig. 5.9. Rates of O_2-consumption and CO_2-release of a diapausing *Cecropia* pupa over an 18-hour period (from Schneiderman and Williams, 1955).

The ventilatory cycle was divided into three stages. Stage I, when the spiracles are open, lasts 15–20 minutes. Stage II, when the spiracles are closed, covers the first hour of the inter-burst period. Stage III, when the spiracles flutter, extends until the next burst occurs, usually 10–15 hours later.

Table 5.2 shows the composition of tracheal gas at the end of each stage. The change in gas composition during each of these stages is also included (Levy and Schneiderman, 1958). Their values were obtained by taking samples from cannulated spiracles. Kanwisher found similar values by vacuum extraction of the entire pupa. This demonstrates the homogeneity of gas composition of the tracheal system and confirms the view that the tracheal gas phase is not an important diffusion barrier. The tracheal volume of a 5 g pupae was found to be about 250 mm^3 by measurements of the compressibility of the submerged pupae.

Fig. 5.10 shows CO_2- and O_2-flux between the pupae and the atmosphere. The data from this figure when combined with those of Table 5.2 give a quantitative picture of gas exchange in a diapausing pupa.

TABLE 5.2

Composition of tracheal gas in the silkworm (*Cecropia*) pupae. Units: percentage of one atmosphere (from Kanwisher, 1966).

Change during I	End of I	Change during II	End of II	Change during III	End of III
O_2 +0·13	0·18	−0·13	0·05	0	0·05
CO_2 −0·03	0·03	+0·01	0·04	+0·02	0·06
N_2 −0·11	0·79	+0·12	0·91	−0·02	0·89

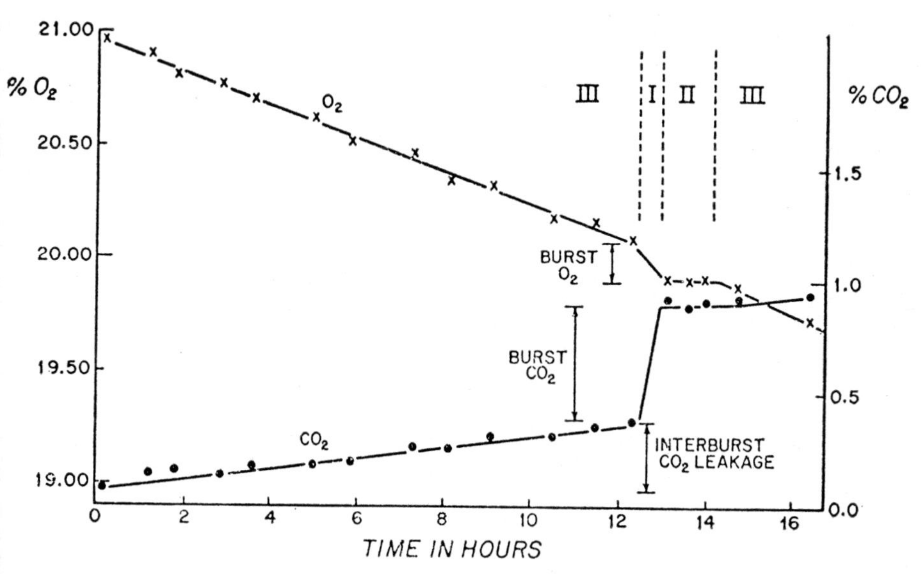

Fig. 5.10. Time change of O_2- and CO_2-concentration in a sealed chamber containing a single *Cecropia* pupa. This method was about twice as sensitive to CO_2 as to H_2O (from Kanwisher, 1966).

During diapause metabolism has an R.Q. of 0·7 and an O_2-consumption of 20 mm^3 O_2/g/h. Thus a 5 g pupae has an O_2-consumption of 100 mm^3 O_2/h and a CO_2-production of 70 mm^3/h.

During Stage I gas and water vapour exchange occurs across the open spiracles by diffusion. Nitrogen enters and the tracheal P_{N_2} is reduced from 89 to 79% atm. Oxygen enters and the tracheal P_{O_2} increases from 5 to 18% atm. The tracheal change in P_{CO_2} from 6 to 3% atm indicates a loss of only 7 mm^3. However, 100 times this is actually lost (measured by change in ambient P_{CO_2}). This means that most of this CO_2 must have been chemically

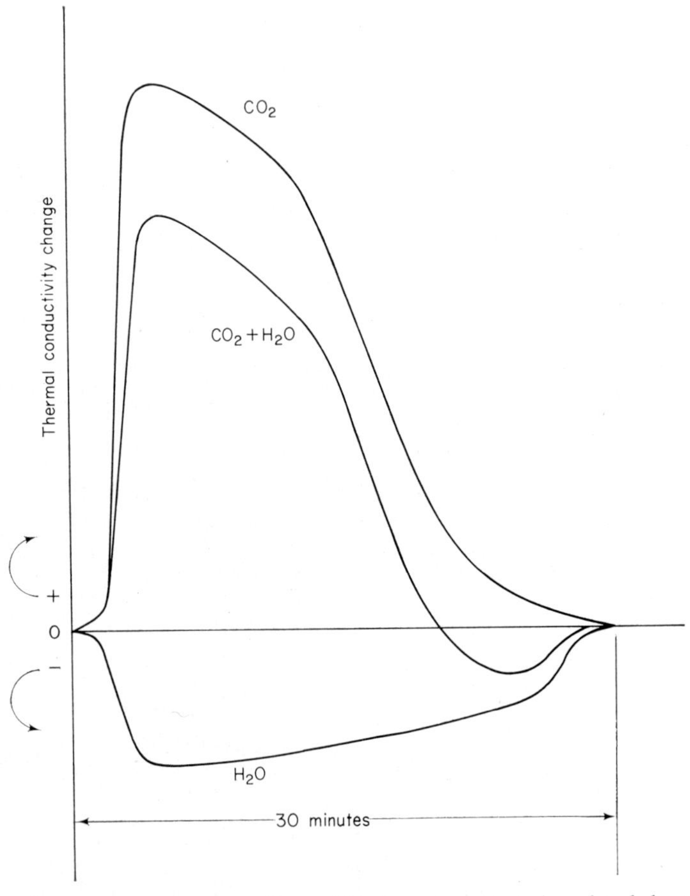

Fig. 5.11. Time course of changes in CO_2 and H_2O vapour content in a closed chamber during one burst of a *Cecropia* pupa (from Kanwisher, 1966).

bound in the tissues prior to the burst. Thus, as CO_2 diffuses from the tracheal gas to the ambient atmosphere, it is replenished from tissue carbonates. Direct measurements of change in ambient humidity (Fig. 5.11) shows that water loss is about half the CO_2 loss.

At the end of this stage the rate of water loss exceeds CO_2 loss. The spiracles may thus appear to stay open somewhat longer than appears necessary. This most likely indicates that the spiracles are kept open to fulfil other needs.

The net result is that during this stage a high tracheal P_{O_2} has been obtained, stored CO_2 is given off while about half as much H_2O as CO_2 is lost.

The spiracles close and Stage II begins and lasts for one hour. During this time the spiracular P_{N_2} increases and N_2 enters the pupa. Tracheal P_{O_2} decreases from 18 to 5% atm CO_2 and water vapour loss during this stage was found to be negligible. The reason for this is most likely that the O_2-consumption establishes a slight underpressure in the tracheal system. This causes air to flow inwards owing to a difference in total pressure, thereby minimizing the outward diffusion of CO_2 and H_2O. The metabolically produced CO_2 is presumably bound as bicarbonate in the tissues, although this has not been experimentally verified.

As the tracheal P_{O_2} approaches 5% atm the spiracles open slightly or flutter. This marks the onset of Stage III which lasts for 15 h. During this time the intracheal P_{O_2} is constant, therefore O_2-consumption by the tissues must be balanced by O_2-flux through the spiracles. The P_{CO_2} increases during this stage from 4 to 6% atm. Thus the ΔP_{O_2} between tracheal gas and the atmosphere is 16% atm, the ΔP_{CO_2} is 5% atm and one would expect the CO_2-diffusion out to be one-third of the O_2-diffusion in. This is confirmed by measurements of composition of the ambient air. Thus about one-third of the metabolically produced CO_2 is released, the rest presumably is bound as bicarbonates in blood and tissues. The water loss was about the same as the CO_2 loss.

The fact that some CO_2 is retained eventually leads to a P_{CO_2} that triggers full opening of the spiracles and Stage I starts again.

The low water loss is thus achieved by hindrance of out-diffusion during Stage II—(owing to the incoming air stream), and by water diffusion parallel to CO_2 loss during Stage III by having the almost sealed spiracle as a diffusion bottleneck.

If the spiracles are kept open during the inter-burst period, the insect is as doomed as a man without water in the desert. Water loss will now continue at the rate it had during the actual burst, i.e. 30 times higher than when the spiracles are functional. Kanwisher (1966) also showed that in a 30% humidity atmosphere pupae with cannulated spiracles will die in a couple of weeks. During this time they had a water loss that equalled 25% of their weight. Normally they lose only a small fraction of this amount.

It is interesting to notice that when the ambient temperature falls below about 10°C no burst occurs. Evidently diffusion through the closed or fluttering spiracles is sufficient for ample gas renewal of the tracheal system at this low metabolic level.

6. ADAPTATION OF THE TRACHEAL SYSTEM OF AQUATIC LIFE—PLASTRON BREATHING

Most aquatic insects also have gas-filled trachea during submersion. Many *Hemiptera* and *Coleoptera* carry a gas bubble or a film of gas under their wings, between legs or between body hairs. The spiracles open into this gas space, the air gill, and gases are exchanged both between the trachea and the gas space and between the gas space and the ambient water.

The principle properties of such air gills have recently been treated in quantitative form by Rahn and Paganelli (1968). Their work is based on the data of Ege (1918) who was the first to describe in detail the properties of air gills and on the work of Thorpe and Crisp, referred to in Crisp (1964) on so-called "plastron" respiration. In the following, we shall briefly review the most important aspects of such systems.

Rahn and Paganelli (1968) distinguish two types of air gills, the compressible type and the incompressible type. The first type is found among water beetles and other insects which periodically come to the surface and renew the gas. In these the gas occurs as a single bubble, held under the wings or between the feet. The pressure inside the bubble is equal to the hydrostatic pressure plus that resulting from the surface forces. The incompressible type consists of gas films held between tightly placed hydrophobic hairs. In the bug *Aphelocheirus* there are one million hairs per mm^2. Each of them measures 0·2 μ in diameter and 5 μ in length. The hairs are sufficiently tightly packed and hydrophobic enough to resist volume changes of the trapped gas when the pressure inside the plastron decreases. The structure and stability of such plastron systems have been analysed in detail by Crisp and Thorpe (see Crisp, 1964).

Rahn and Paganelli (1968) describe the behaviour of both types of such air gills. The main property of both is that the gas phase provides a micromilieu across which gases exchange. The gas space has a much larger contact area with ambient water than the spiracular opening. In both cases the inert gas, N_2, plays an essential role. In the compressible type the bubble will shrink when the insects use O_2. The expired CO_2 will play a minor role since, owing to its high solubility in water, it will escape from the bubble. The shrinking will cause increased P_{N_2} in the bubble. The result is that N_2 diffuses out of and O_2 into the bubble from the water. On the assumption that the contact area between the bubble and the water remains constant, Rahn and Paganelli (1968) showed that after some time the composition of the bubble reaches a stable plateau. However, the bubble is continuously shrinking since N_2 continues to diffuse outward.

The time it takes for the bubble to disappear completely varies with the depth since the hydrostatic pressure, about 1 atm per 10 m, causes the P_{N_2} to increase above the ambient. Fig. 5.12 gives an idea of the lifetime of this type of compressible air gill at different depths.

The animal is supplied by O_2 from two sources (1) the O_2 initially present in the air gill and (2) the O_2 that diffuses into the air gill from the ambient water. The second factor is often the most important. Ege (1918) found that in *Notonecta* it may contribute up to 13 times more O_2 than that present in the bubble initially.

The second type of air gill, the incompressible type or plastron is distinguished by a permanent gas space. The permanence depends on the constant volume which means that neither hydrostatic pressure nor O_2-uptake will

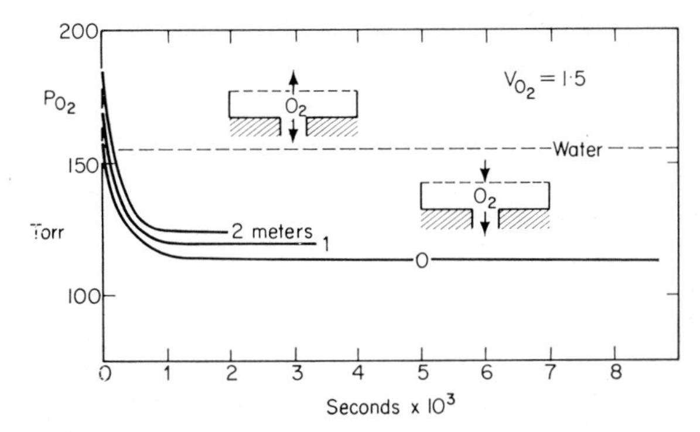

Fig. 5.12. The lifetime and P_{O_2} of a model gas gill initially containing air and placed 0, 1 or 2 m below the surface. The insets show directions of net O_2 movements, P_{O_2}s above the dashed line (in the unsteady state), mean loss of O_2 to water; below the line, O_2 diffuses into the gill. Note the shortening of the gill lifetime with depth (from Rahn and Paganelli, 1968).

change P_{N_2} which will therefore be equal to the ambient P_{N_2}. Consequently there is no net transport of N_2. Also in this case the P_{O_2} will show an initial fall before it is stabilized at a value where O_2-uptake of the insect is balanced by O_2-diffusion from the ambient water.

Paganelli *et al.* (1967) made a working model of such a system. The incompressible air space was made from a glass jar with a millipore filter lid. This unit was immersed in water. Inside the jar was an animal. In this model they found no ΔP_{N_2}, and a ΔP_{CO_2} about one-twentieth of the ΔP_{O_2} from gas to water. The P_{O_2} in the jar depended upon the O_2-uptake of the captive animal and on the permeability and dimensions of the millipore exchange membrane.

Thus the compressible type air gill is most suitable near the surface and for the insects which do not mind periodic refuelling trips to the surface. The plastron type air gill acts equally well at any depth (within limits, Crisp, 1964) and will be suitable for any insect which lives permanently below water. I have tried in vain to find a rational reason why not all insects have plastron gills. My answer is that the bubble gill is a good excuse to visit the surface. Even this explanation is a poor one, however, since very few insects appear to use the opportunity for anything else than a renewal of the bubble. Some break the surface only with their rear ends, others with feet or feelers or special tubes. Some even puncture underwater air-filled plants.

In an aquatic elmid beetle, Stride (1958) reported a remarkable and completely different device for maintaining a permanent physical gill. This beetle (which can fly) lives on the surfaces of stones submerged in torrents and carries a very large air bubble extending from the first femora back over and past the dorsum. The bubble forms because the bladelike flattened femora, and vanes on the pronotum, deflect the water up and away from the elytra. In calm water the bubble is gradually lost and the animal dies if prevented from surfacing, even though the water is saturated with air. Stride's explanation of the bubble's utility and permanence is that in currents faster than 70 cm/sec the Bernouilli-Venturi effect of flow across the convex surface of the bubble causes the interior to be at a total pressure that may be several centimetres of water below atmospheric. Provided the ambient water is nearly saturated O_2 will enter the gas phase.

A similar case of replenishment of gill air independent of external gas sources was described by Hinton (1953). In the stationary *Oxycera* larvae, body contractions drive a bubble of gas out of the post-abdominal spiracle, where it is held by a fringe of setae. The post-spiracular chamber is then closed and the body-wall muscles relax. Since felt prevents water from entering the anterior spiracles, an intratracheal vacuum develops and gas diffuses in from the tissues, being replaced by gas from the ambient water entering the body through the permeable cuticle. By removing bubbles as they appeared, Hinton showed that the submerged larva can form new ones to a total of more than the entire volume of the body. Possibly the same mechanism is used by the *Ephydra* larva, in which Nemenz (1960) suggested that bubbles protruding from the posterior spiracles serve as a gill.

Many insect eggs have areas which are penetrated by a complicated meshwork of air channels, the diameter of the meshes is well below 1 μ and they are lined by hydrophobic material. These porous areas of the egg shells undoubtedly act as a plastron and constitute the normal route of gas exchange between the developing embryo and the environment. The detailed structure and functional significance of several insect egg plastrons have been described in several papers by Hinton (see Crisp, 1964).

When submerged in water these areas act as the plastrons of aquatic insects although the gas phase in many instances appears to be separated from the tracheal gas by a thin membrane. When exposed to air they function in much the same way as the shell of bird eggs. The gas-filled channels of both shells act as a microrespiratory environment for the developing embryo. This arrangement has an advantageous effect on the ratio of O_2-uptake to water loss, since the exchange of both will be determined by the rate of diffusion through the gas phase. The diffusion coefficient of O_2 in air is about two-thirds that for H_2O vapour. However, since the partial pressure difference across the system will approach 100 mm Hg for O_2, but rarely exceed 30 mm Hg for H_2O, sufficient O_2-uptake can occur without excessive water loss. Many insect eggs, particularly from species developing in a hot and dry environment, have a thin external hydrophobic cuticle. A similar arrangement is found in bird eggs.

The plastron structure has a parallel in the fine hydrophobic air channels of many plants which prevents the entrance of water. This property is the main reason why ancient Egyptians and modern Norwegians have had success when they attempted to stay afloat for months on board a ship of straws sailing across the Atlantic Ocean.

C. Respiration by Lungs and a Circulatory System with Respiratory Pigment

The principal difference between tracheal and lung respiration is that in the latter a circulatory system transports O_2 and CO_2 between the respiratory organ and the tissues. The properties of this transport mechanism are therefore of special importance.

From a functional point of view lungs can be divided in two categories: diffusion lungs and ventilation lungs. In the former renewal of gas occurs by diffusion. This type is the more primitive of the two, found in molluscs and arachnoids, while ventilation lungs are found in larger molluscs and in all vertebrates. The fact that diffusion is efficient only over short distances is reflected by the observation that lungs of all larger animals, and of small animals during activity, are ventilated. This is a parallel to the appearance of ventilation of tracheal systems already described.

1. Diffusion Lungs

The distinction between diffusion lungs and ventilation lungs is not sharp and only of limited interest. The important point is that in small animals with low metabolism gas renewal of the lungs may occur without ventilation. In most animals with such lungs ventilatory movements are observed during activity. From a structural point of view we can distinguish one group of diffusion lungs as *book lungs*. These are found primarily among spiders, where they occur as paired organs on the abdomen. The book lung consists of a series of blood-filled, parallel, thin plates separated by thin layers of air. In the giant South American tarantula each "sheet" of the lung is 10 μ thick with a wall thickness of less than 1 μ. The sheets are separated by 5 μ air spaces (personal observations, Fig. 5.13). These dimensions suggest a very specialized organ. The book lung is enclosed in an invagination of the abdominal wall and renewal of the air does not appear to involve ventilation.

The air space is guarded by a spiracle. Although it does not seem to have been investigated in detail the same compromise between gas exchange and water loss seems to occur here as in insects.

Hazelhoff (1926) and Fraenkel (1930) showed that in a number of spiders and a scorpion the spiracles remain almost closed when the animal is resting and the temperature is not too high. During activity the spiracles open periodically. Hazelhoff showed that CO_2 controlled the opening of the spiracles. In 1% CO_2 the spiracles open to one-tenth of the maximum while in 2·5% CO_2 to one-half its maximum opening. Each spiracle was shown to react individually to the local P_{CO_2}. Low P_{O_2}, below 5%, also causes the spiracles to open.

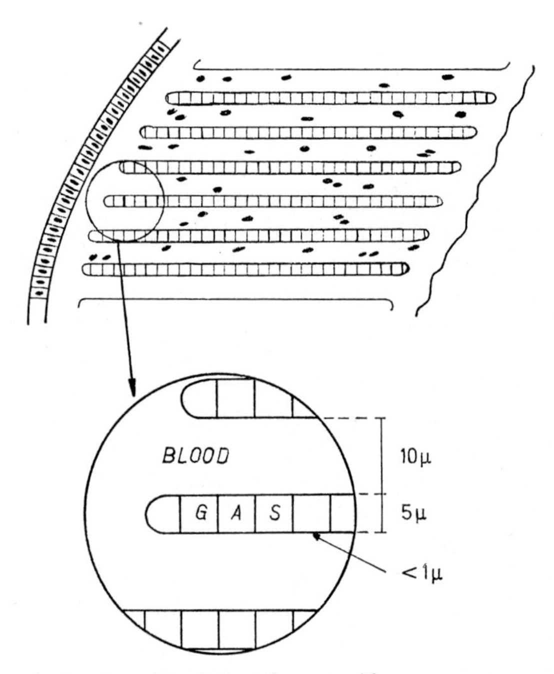

Fig. 5.13. Schematic drawing of book lung from a spider.

So far, no one has attempted a physiological description of gas exchange by book lungs. We know neither the P_{O_2} or P_{CO_2} inside the lung, the gas gradients across the respiratory surfaces, nor the degree of O_2-extraction.

Apart from the described book lung all other lungs are of one type, namely invaginations of large membranes into the body. The most primitive lungs are found in snails and slugs. In these animals the mantle cavity is developed into a lung. Its surface is increased by a number of ridges and is well vascularized. In many ways it resembles a smaller version of the mouth cavity found in the electric eel. In these animals the lung is a far less specialized organ than in higher animals. This is true both with regard to its anatomy and to its function. Apart from gas exchange, the lung serves as a hydrostatic organ and prevents evaporation of water, which in many cases is important in temperature regulation. Krogh (1941) found that the volume of the lung in *Helix pomatia* was 5–7 ml when the animal was out of the shell, but only 0·5 ml when it was retracted. This indicates that the lung of shell-bearing species takes part in the mechanism of retraction of the animal into the shell. In the naked slugs the volume is smaller, about 0·3 ml in an *Arion* that weighs 10 g. In this animal Dahr (1927) found that the respiratory surface area measured 6–7 cm^2. According to his calculations the partial pressure difference necessary to supply the lung surface with oxygen by diffusion from the outside, amounted to 2 mm Hg. This shows that even in considerably larger forms and at low oxygen pressures, diffusion of oxygen and CO_2 inside the lung will be ample. Thus nothing could be gained by respiratory movements with regard to gas renewal. Nevertheless ventilatory movements have been described several times (Ghiretti, 1966a) and most likely they have some functional significance.

Dahr (1924) described ventilation and pneumostome behaviour in *Arion* and *Helix*. In moist air with normal oxygen pressure the pneumostome only opens at intervals. Dahr found that at 15 mm P_{O_2} *Arion* will close its pneumostome 15–30 times in 30 min for about 7 min in all. During the closed period the enclosed air is moved by muscular contractions. The resultant pressure increase inside the lung, which amounts to about 20–30 mm H_2O, is too small to be of any importance for gas exchange. More likely the muscular movements of the lung walls when the pneumostome is closed cause turbulence of the liquid lining. Thus better contact is made between the gas and respiratory tissue or possibly the lung circulation is improved.

Maas (1939) found that the frequency of ventilation in air-breathing gastropods depended on *both* CO_2, O_2, humidity and temperature. This observation further strengthens the notion that in these animals ventilation is primarily concerned with regulating water loss.

In fresh-water pulmonates Precht (1939) found that ventilation was regulated by at least two stimuli. One is the oxygen tension, the other the amount of air present in the lung. The latter is due to the fact that the lung also serves as a hydrostatic organ. If such a snail lives in water above which is an atmosphere of pure nitrogen, it will remain at the surface with the pneumostome open for a couple of minutes, and then go down, only to return very soon. If the animal is exposed to an atmosphere of pure oxygen, it will remain the same time at the surface, but will return to the surface long before the oxygen supply in the lungs is exhausted, namely when the *volume* of the lung has reached a certain value. The animal can be forced to return to the surface by increasing the ambient pressure, i.e. reducing the volume. CO_2 in low concentrations had very little influence upon the ventilatory frequency. This is thus one of the few cases of an air-breathing animal being rather insensitive to CO_2. This case, like all the others that have been mentioned, is associated with accessory aquatic gas exchange. Most likely it depends on the fact that part of the CO_2 produced under normal conditions is lost by ways other than the lung (for example, by direct diffusion through the body surface).

In several small, but unrelated animals, the diffusion lung has a form that combines the features of gastropod lungs with the tracheal system of insects. Spiracles open from the surface into a large vestibulum, from which hundreds of short air tubes originate to form a three-dimensional tube system. Blood circulation around these tracheae is ample and carries gases to and from the tissues. In principle this system acts like a lung, the trachea being analogous to alveoli. This arrangement is found in some naked, tropical snails (*Janellidae*), in the isopod *Porcellio* (sow bug) and in the chilopod *Scutigera* (house centipede) (Krogh, 1941).

2. VENTILATION LUNGS—GENERAL PROPERTIES

The structure of ventilation lungs shows a development from the simple uni-alveolar lung of the lungfish *Neoceratodus* and some amphibia, to the highly organized multi-alveolar lung of homeothermic animals. Paralleling this increase in specialization is an increasing number of functional properties and problems. In general we find, as was the case with the development of gills, that the more specialized a lung is, the more adapted it is for diffusion with regard to all the pertinent parameters.

Among water-breathing animals even primitive species have respiratory organs which extract a large part of the O_2 present in the inhaled water. Thus the respiratory efficiency is high. In air-breathing animals this is not the case. Very rarely is more than 25% of the inhaled O_2 utilized. This is reasonable for several reasons (Ch. 2). One is that the energy necessary to move air is much smaller than that necessary to move water. The demand for a high utilization is therefore not critical. Another is a consequence of two-way ventilation: the decrease in O_2 content is approximately equalled by an increase in the partial pressure of CO_2. The rate of ventilation in air-breathing animals is regulated primarily by the P_{CO_2} in the lung, and not as we found in water-breathing animals, by the P_{O_2}.

The constancy of the aerial environment compared to the aquatic is reflected in the smaller variability of the respiratory properties of blood from air breathers than from aquabreathers. However, three types of hypoxic conditions are accompanied by marked changes in the blood. These are the permanent exposure to hypoxia at high altitudes, the temporary hypoxia during diving and the transitional one during foetal life. These subjects will be treated separately. Here I shall only outline the general properties of blood from lung breathers.

The blood contains Hb of molecular weight about 68 000 contained in cells. Each molecule can bind four molecules of O_2, thus the O_2-capacity is

1·34 ml O_2 per g Hb. The red blood corpuscles, erythrocytes, are of varying size and shape and in higher vertebrates, often disc shaped with a diameter of 5–10 μ. The mammalian erythrocyte has no nucleus, thus it is more a package of Hb than a proper cell. Erythrocytes of reptiles, amphibians and birds are nucleated.

The O_2-capacity of whole blood is determined by the Hb content. Within the same species the Hb content per cell is constant and the O_2-capacity is therefore proportional to the number of erythrocytes per unit volume. This is frequently expressed as the volume per cent of red cells in the blood or haematocrit.

TABLE 5.3
The magnitude of the Bohr-shift in some animals expressed as the change in P_{50} per unit change in pH (from Prosser and Brown, 1962).

	$\Delta \log P_{50}/\Delta pH$		
Mouse	−0·96	Horse haemoglobin	−0·68
Guinea-pig	−0·79	Crocodile	−0·8
Rabbit	−0·75	Bullfrog	−0·24
Dog	−0·65	Mackerel	−1·2
Man	−0·62	Lamprey (*Petromyzon*)	−0·7
Pig	−0·57	*Polistotrema*	−0·0
Cow	−0·52	*Arenicola*	−0·0
Elephant	−0·38	*Gastrophilus*	−0·0
Duck	−0·67	*Lumbricus*	−0·25
Surf scoter	−0·58	*Eupolymnia*	−0·0
Western grebe	−0·45	*Spirographis*	−0·66
Horse myoglobin	−0·10	(chlorocruorin)	

The O_2-capacity of blood from amphibians and reptiles is from 5 to 12 volume per cent, while in birds and mammals it is from 15 to 20 volume per cent if we exclude animals adapted to hypoxic conditions (Prosser and Brown, 1962). To appreciate the physiological significance of variations in O_2-capacity we should keep in mind that under normal resting conditions only one-third to one-fourth of the O_2-capacity is utilized (expressed by the A–V difference). Only during high metabolism does the A–V difference approach the O_2-capacity. Venous blood from a working muscle may in fact be almost devoid of O_2. Thus a high O_2-capacity would be expected to occur in animals which can greatly increase their O_2-consumption. I am not aware of direct evidence on this point, but it appears likely that homeotherms have greater ability to increase their metabolism than poikilotherms.

Physical training, which increases the maximal O_2-uptake appreciably does not involve an increase in blood O_2-capacity but rather an increased cardio-vascular capacity.

No correlation is found between the adult size of animals and blood O_2-capacity.

The O_2-dissociation curve of vertebrate blood is of sigmoid shape. At normal environmental temperature and blood pH, the P_{50} of amphibian and reptilian blood is between 12 and 20 mm Hg; in mammals at body temperature and normal P_{CO_2}, it is from 20 to 60 mm Hg; while in birds it varies from 35 to 60 mm Hg (Prosser and Brown, 1962). The Bohr-shift is remarkably similar among homeotherms (Table 5.3) when compared to the variability among fishes. This appears to be a reflection of the constancy of the respiratory environment.

D. The Alveolar Lung

1. STRUCTURE

The most primitive vertebrate lungs are found in urodeles like *Proteus*. Their lungs consist of a single smooth-walled sack containing a dense network of blood capillaries in the wall. The gas to blood distance is about 4 μ. A slightly more advanced type is found in lungfishes and amphibia where the surface is folded and the surface to volume ratio therefore increased. This trend is further developed in amphibia like frogs and toads where also secondary ridges are often present.

Czopek (1966) reviewed information on the vascular anatomy of the amphibian lung. He expressed the structural adaptation for gas exchange by a coefficient obtained by dividing the respiratory area (the area that is capillarized) by the total external surface area of the lung. This coefficient is less than one in amphibia with primitive lungs (e.g. in the genus *Bombina*). In the frog *Rana esculenta* the coefficient equals 8. Czopek has also counted the number of capillary meshes per mm^2 in a large number of amphibians. In the vestigal lung of *Rana olympicus* he found 74 meshes/mm^2, while in some urodeles up to 550, and up to 1000 in the anuran *Scaphiopus couchi*. Czopek has also measured/calculated the ratio of the total capillary length to the body weight. In urodeles it is 2·3 to 5·8 meshes/g, in anurans from 3·4 to 34·4 m/g. In urodeles similar ratios for the skin vascularization exceed those for the lungs, while in most anurans the reverse holds true.

Many turtles have lungs that are more specialized; the degree of subdivision is increased and the blood to gas distance approaches 1 μ. The specialization of the lung as a respiratory organ is further developed in mammals. There exists extensive literature on the respiratory function of mammalian lungs; in this context only an outline of our knowledge can be presented. A quantitative account of the morphological features that are of functional importance is given by Weibel (1963) and a very clear description of the organization of the lung by Staub (1963).

The mammalian lung is characterized by a very high ratio of alveolar

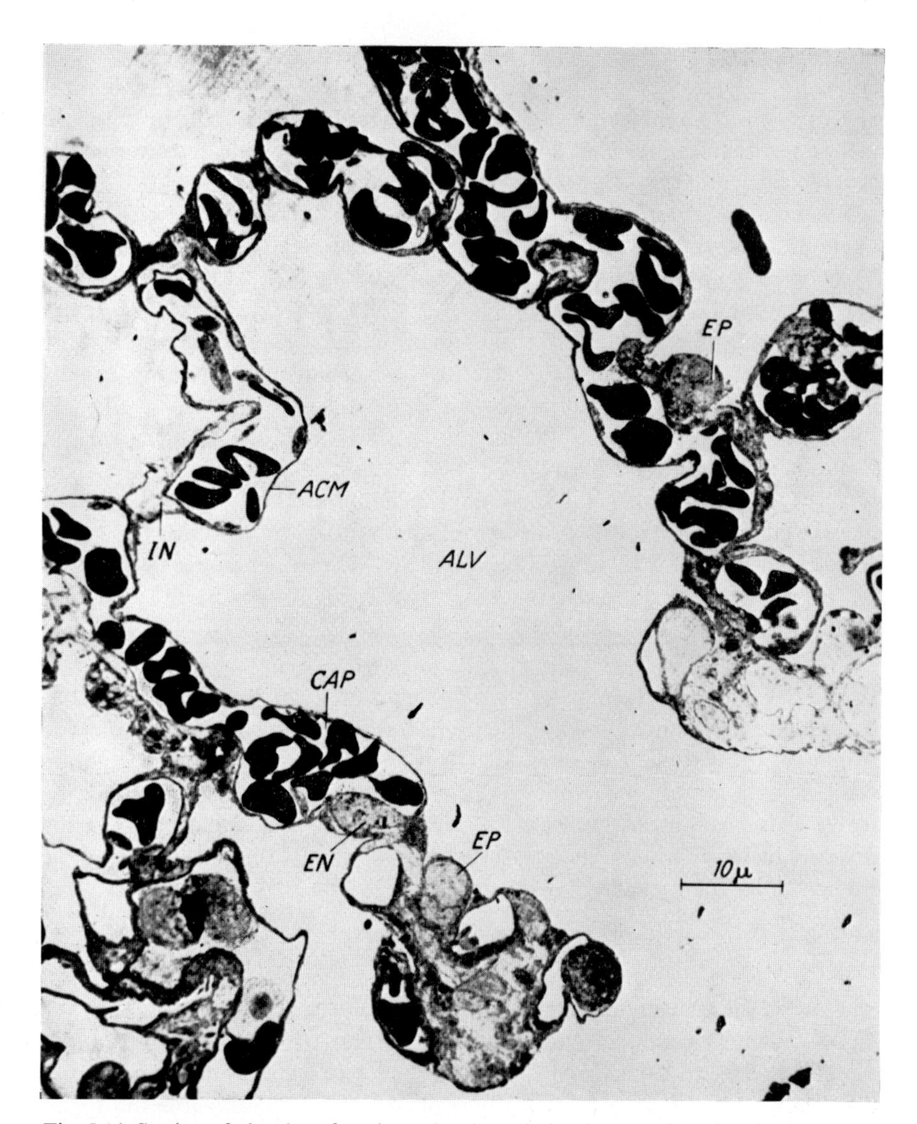

Fig. 5.14. Section of alveolus of rat lung showing relation between blood and alveolar gas. The alveolo-capillary tissue is very thin over large areas (ACH) but is thicker where it contains alveolar epithelial (EP) and capillary endothelial cells (EN) or interstitial elements (IN). ×1350 (from Weibel, 1963).

surface to lung volume. A 5-litre human lung contains 75 m^2 of alveolar surface (Weibel, 1963). The large respiratory surface is obtained by subdivision of the surface into spherical alveoli (about 300×10^6 in a human lung) which are arranged relative to each other, much like the bubbles of froth. The respiratory capillaries form a meshwork on the alveolar surface and due to the geometry most capillaries are exposed simultaneously to two neighbouring alveoli (Fig. 5.14). About 75% of the alveolar surface is covered by capillaries. The gas to blood distance is less than 1 μ.

Several alveoli make up an air sac, and these are interconnected by alveolar ducts and connected to the airway system by bronchioles. The number of branchings from alveoli to the trachea ranges from 7 to 24 depending upon the position of the alveoli.

The respiratory surfaces are supplied by deoxygenated blood via the pulmonary artery. The degree of arteriovenous shunts in the lungs is usually below 5% of the total blood flow. The passage time of blood through the pulmonary capillaries of the human lung has been estimated by various methods to be about 0·75 sec at rest, and 0·30 sec at maximal exercise. This time appears fully sufficient for complete equilibration of gases, but it is questionable whether the more complicated processes like the Bohr-shift with a half time of 0·12 sec will be completed. Direct experimental evidence on this point is difficult to obtain owing to the small dimensions of capillaries.

2. STABILITY OF THE ALVEOLI

The development from a simple bladder to one which is subdivided into minute and mechanically weak spheres also involves a problem of geometric stability. The problem is illustrated by a soap foam: the small bubbles coalesce with larger ones, until finally a few large bubbles have replaced the froth. This phenomenon reflects the action of the surface tension upon the pressure inside a bubble. The pressure difference across the wall of the bubble P is given by the formula

$$P = \frac{2\pi}{r}$$

where π is the surface tension and r the radius. The smaller the bubble, the higher its internal pressure as long as the surface tension is constant. Since the alveoli are gas pockets with little mechanical rigidity, one would expect their fluid lining to cause the same type of coalescence observed in the foam. This would rapidly cause many small alveoli to coalesce into a few large ones, which would be detrimental to their function. In fact one would arrive back at the primitive type of lung. A development of this type is seen in many mammalian infants dying of a certain type of respiratory distress.

The surface tension of the fluid lining the lung also affects other aspects of the lung function. If the surface tension is high, then a higher pressure is needed to stretch the alveolar surfaces than if it is low, i.e. the cost of breathing increases. The surface tension will also affect the fluid balance across the alveolar membrane. The higher the surface tension the higher will be the "suction" force acting together with the capillary blood pressure and subatmospheric thoracic pressure to cause filtration of fluid from the capillaries. Ideally, therefore, the fluid lining of the alveoli ought to have a low surface tension which varies with the distension so that the pressure inside alveoli of varying size stays constant. This is in fact what has been found to be the case.

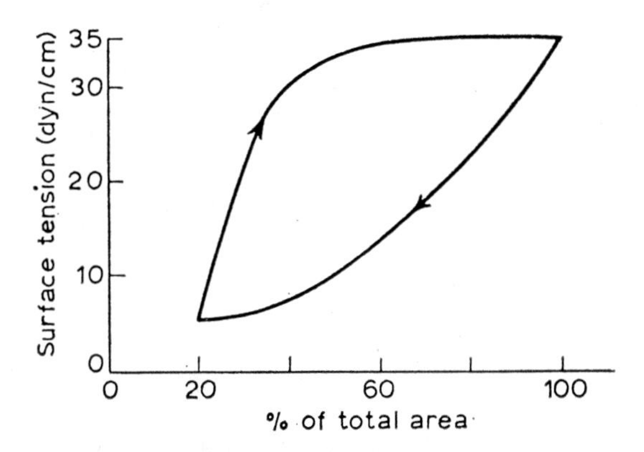

Fig. 5.15. Course of variation of surface tension of lung extracts when the area is varied cyclically on the Wilhelmy trough (schematic only; not the result of an actual experiment) (from Pattle, 1966).

The properties of the surface fluid coating the lungs have been ably reviewed by Pattle (1966) and by Clements and Tierney (1964). Plasma has a surface tension of about 30 dynes/cm. When the surface tension of washings from the lungs or of extracts of the minced lung is measured the values obtained depend on the conditions of the experiment. Sutnick and Soloff (1962) found that if the surface of such fluids was compressed and the surface area then kept constant, the surface tension rose from 9 to 30 dynes/cm over a period of 90 min. This shows that such a fluid surface is not in equilibrium immediately after a change of area, but will in time reach its state of minimum free energy. Fig. 5.15 shows a schematic graph of the surface tension measured during a cyclical variation in the area of lung extracts on a Wilhelmi trough. The essential feature is that the surface tension decreases when the area is compressed and increases when it is stretched. The hysteresis reflects the aforementioned nonequilibrium state of such surface films.

Thus these measurements demonstrate the two main properties which appear desirable to resist transuduation and create equal pressures in alveoli of varying size and in various stages of distension. Clements *et al.* (1961) developed a mathematical model of the stability of the alveoli based on the notion that it was due only to the properties of the surface material. When experimentally determined values for the surface properties and alveolar dimensions were inserted, the model predicted that the alveoli would live in stable coexistence. This success of the model indicates that the surfactant of the lungs is indeed an important stabilizer.

The surface properties of the fluid lining of primitive lungs have also been measured. The lung fluids from lungfishes (*Protopterus*), amphibians and reptiles all show principally the same features (Pattle and Hopkinson, 1963).

This surface film consists of lipoprotein, which is rich in phospholipids and is probably present *in situ* as a jelly. It is readily permeable to gases (Pattle, 1966).

3. VENTILATION AND RENEWAL OF GAS

Two types of ventilation are found. The primitive condition is found among amphibians which inspire by swallowing air present in the mouth cavity. Expiration is passively achieved by opening the glottis. The details of amphibian ventilation are intricate and well reviewed by Foxon (1964). In most reptiles and in all mammals, inspiration occurs by suction. The lungs are placed in a pleural cavity inside the thorax which, upon expansion, creates a sub-atmospheric pressure inside the lung which causes the inflow of air. Expiration is predominantly passive although active expiration may occur during stressful conditions. By ventilation part of the gas in the lungs is replaced by air, primarily that contained in the larger airways. Renewal of gas at the respiratory surfaces occurs mostly by diffusion. Little doubt exists that diffusion is sufficient for renewal of alveolar gas. This has the consequence that the composition of the alveolar gas fluctuates less than if the alveoli had been directly ventilated with air.

In many animals ventilation also serves as the regulation of body temperature. Such ventilation, called panting, involves mainly the upper airways. Even during strong panting the respiratory units are not hyperventilated and hypocarbia or alkalosis do not occur in unanaesthetized dogs to any significant extent (Hemingway and Barbour, 1938; Albers, 1961).

4. REGULATION OF CIRCULATION AND VENTILATION

Pulmonary circulation and ventilation are accurately regulated both in relation to each other and to the functional state of the individual. Thus in man the composition of expiratory gas and of arterial blood is about the same at rest and during exercise and the ventilation volume increases in direct proportion to the O_2-uptake. The cardiac output does not increase in exact proportion to O_2-uptake since the venous saturation is less during exercise than during rest (Fig. 5.16).

The regulatory mechanisms behind this precise co-ordination are known only in qualitative terms. That is to say: we know many of the stimuli involved and their action on ventilation and circulation but not the quantitative contribution of each under various conditions. The respiratory function is regulated on two levels. One is total circulation and ventilation, the other the distribution of gas and blood to the various parts of the lung. The total circulation and ventilation is regulated partly by chemical and partly by mechanical stimuli. The chemical composition of arterial blood, particularly the P_{O_2}, P_{CO_2} and/or pH is registered by peripheral chemoreceptors (carotid and aortic bodies) and by central receptors. These latter are also influenced by the composition of the cerebrospinal fluid. Impulses from these receptors are conveyed to the respiratory centre whose activity controls ventilation.

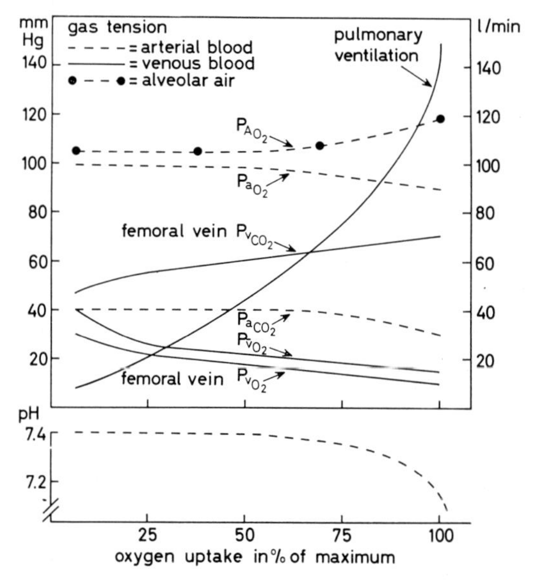

Fig. 5.16. The relationship between some of the important respiratory parameters and O_2-uptake during exercise. The O_2-uptake was varied by treadmill work (graph kindly supplied by L. Harmansen, Institute of Work Physiology, Oslo).

Information about the degree of exercise is also conveyed via anatomically unidentified receptors in the extremities (Kao, 1963). Recent studies (Biscoe and Purves, 1967) indicate that this information goes to the carotid bodies where it modifies the signal to the respiratory centre.

The other type of regulation occurs on the alveolar level where the ratio of perfusion to circulation is of crucial importance for optimal gas exchange. The two processes are regulated by smooth muscle contractions of the arterioles and smaller bronchioles. It has been shown that insufficient ventilation of an area of the lung elicits vasoconstriction in that section via local hypoxia.

5. RESPIRATORY PROPERTIES RELATED TO THE SHAPE AND SIZE OF THE BODY

Certain fundamental aspects of respiratory mechanisms become apparent when we compare the lung function of animals of varying size and shape. Problems of this kind have rarely been thoroughly investigated. The following treatment will therefore be very speculative, but still useful I think. It may suggest problems which, when solved, will augment our understanding of respiratory design.

From the point of view of lung function the length of the trachea is of interest since it is an important component of the dead space and ventilation resistance. Do, for example, the giraffe or the heron display anatomical or

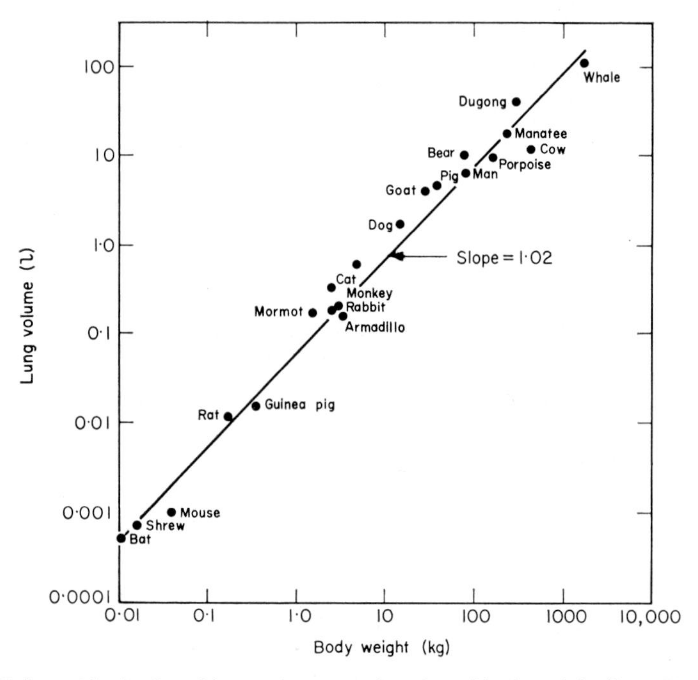

Fig. 5.17. Logarithmic plot of lung volume as a function of body-weight (from Tenney and Remmers, 1963).

physiological characteristics which can be connected to their longneckedness? Goode *et al.* (1969) have recently investigated the respiratory effects of breathing through a tube. They found increased end-tidal P_{CO_2}, decreased end-tidal P_{O_2} and increased ventilation. It would be interesting to know if habitual tube-breathers display the same pattern.

Variations of the lung dimensions with body size were studied by Tenney and Remmers (1963). They showed that there exists a linear relationship between body size and lung volume over a wide range (Fig. 5.17). The lungs make up about 10–15% of the total body volume. The resting O_2-uptake per g body weight does not increase linearly but with the two-thirds power of the weight. Thus a mouse has about 15 times higher metabolism per g than a man. This means that a mouse lung must be 15 times more efficient per volume than a human lung. Comparative data on maximum O_2-uptake of various-sized animals are more scarce. Krogh (1941) gives values which show that the O_2-uptake of white mice increases 10-fold with running, while that of man may increase 20-fold. However, the laboratory mouse is even more physically degenerate than man, and it is most likely that wild and active small animals are able to increase their O_2-uptake 20-fold. It seems fair to conclude therefore that about 15 times more O_2 is taken

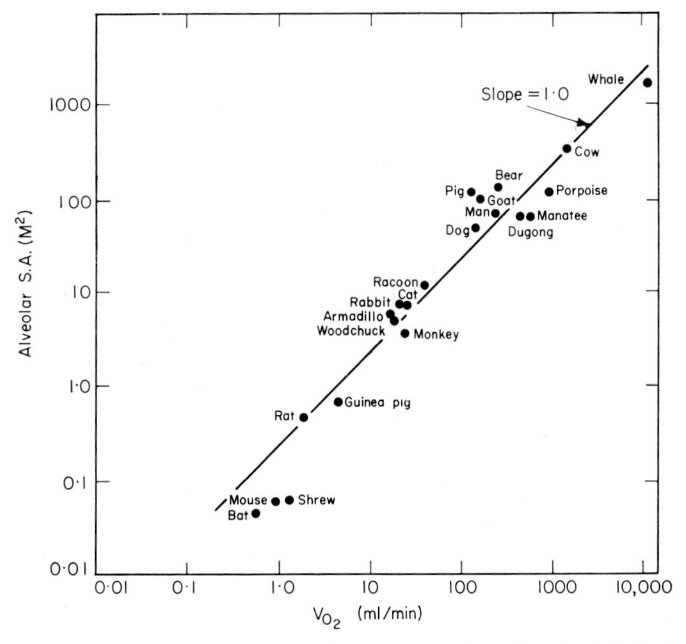

Fig. 5.18. Logarithmic plot of alveolar surface area as a function of whole body oxygen consumption (from Tenney and Remmers, 1963).

up per volume in the mouse lung than per volume of the human lung. If O_2-uptake in both cases is diffusion limited, then this difference must be due to changes in some or all of the diffusion parameters: the area, the diffusion distance, the difference in P_{O_2} between blood and alveolar gas or the permeability of the barrier. Since nothing is known about the permeability we assume that this parameter is constant and independent of body size. A similar scarcity of information prevails with regard to comparative data on the thickness of the diffusion barrier. The ΔP_{O_2} is well known only in man. However, the following data roughly describes the situation. The P_{50} for mouse blood is 50 mm Hg compared to 27 for human blood (Schmidt-

Nielsen and Larimer, 1958). Assuming about 95% saturation of arterial blood for both animals, it is most unlikely that the pulmonary gradient should be higher in the mouse than in man. In fact it appears more reasonable that ΔP_{O_2} is lower in the mouse than in man.

Comparative data on the respiratory surface and other pertinent lung dimensions are fortunately sufficient to give at least a general answer to our question. Tenney and Remmers (1963) showed that the surface area of the lungs is directly proportional to the O_2-uptake of the animal (Fig. 5.18).

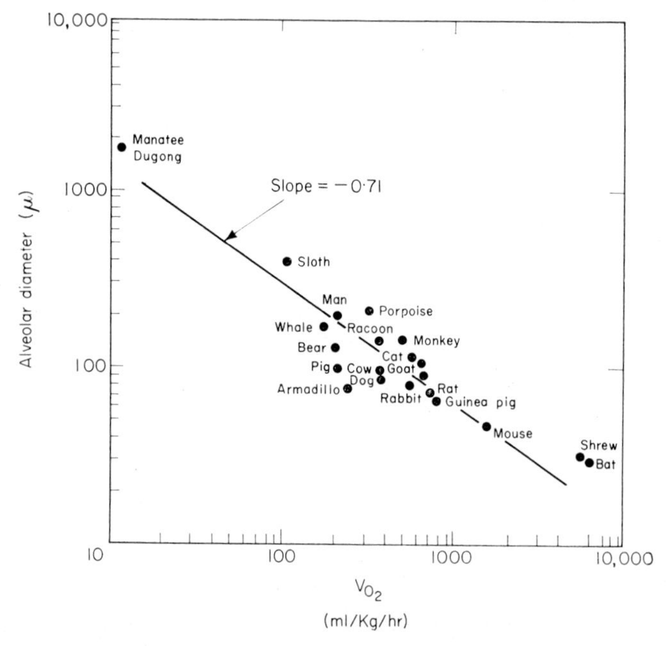

Fig. 5.19. Logarithmic plot of mean alveolar diameter in microns as a function of metabolic rate per unit of body weight (from Tenney and Remmers, 1963).

This correlation supports our previous assumption that there exist only slight variations in the other three parameters. We can conclude therefore that the greater gas exchange per volume lung of mice than of humans is due to a proportional increase in diffusion area per volume of lung. Tenney and Remmers (1963) showed, as was expected, that this increased surface to volume ratio was at least in part due to increased septation of the lung (Fig. 5.19). Thus, whereas the alveolar diameter in man is about 170 μ, in the mouse it is 50 μ. This means that a certain "volume of alveoli" has about 3 times larger surface in mice lungs than in human lungs. A further reduction in the area to volume ratio of the lungs, the "structural efficiency", appears to be an inevitable consequence of the increased size. To illustrate this we can try to make a human lung from a number of mice lungs; to do so the mice lungs must be connected by wide bore tubes, which must again be connected, and so on. All these connecting tubes add volume, but very little respiratory surface. This increase of dead space relative to the alveolar volume has consequences for the efficiency of renewal of alveolar gas. The larger the volume of the lung, the longer the average distance from the respiratory surface to the parts of the bronchial tree which are supplied by fresh air at each breath. Thus the diffusion distance of the gas phase of the lung increases. The increase in dimension of the alveoli with size (i.e. an increase of cross-sectional area) may be considered as a compensation for this. The structural changes of the lung which accompany increasing size will therefore tend to make renewal of alveolar gas more complete, the smaller the animal. This should be reflected by higher alveolar P_{O_2} and lower P_{CO_2} of small than of large animals. Direct data is not on hand, but indirect evidence supports such a conclusion.

First of all the arterial P_{CO_2} of cats is about 28 mm Hg as against 40 for man. I have been unable to find such data for rats or mice although I am certain they exist. Schmidt-Nielsen and Larimer (1958) found the P_{50} of the blood at a P_{CO_2} of 40 mm Hg to decrease with increasing size (Fig. 5.20). Mouse blood had a P_{50} slightly above 50 mm Hg. If this P_{CO_2} is normal then arterial blood should not become 95% before the P_{O_2} approaches 150 mm Hg. This is unreasonable. It appears more likely that the high P_{50} is due to a P_{CO_2} which is abnormally high for the mouse. Riggs (1959–60) determined the magnitude of the Bohr-shift of Hb from animals of varying size. He noticed that although the extent of the shift was larger the smaller the animal, the P_{50} at a pH of 7·4 was almost identical (Fig. 5.21). This suggests that at the normal mammalian arterial pH the P_{50} is size independent, but that the arterial P_{CO_2} at pH of 7·4 is lower the smaller the animal. This supports my earlier suggestion that the alveolar P_{CO_2} is lower in smaller than in large animals, i.e. it indicates more efficient alveolar ventilation.

The fact that the blood P_{50} at pH 7·4 is independent of size might indicate that the ΔP_{O_2} across the lung barrier was also similar. However, Riggs's (1959–60) observation of a larger Bohr-shift in smaller than in larger animals will influence the situation. It means that the smaller the animal the larger will be the ΔP_{O_2} experienced by the blood during its passage through the lungs: as CO_2 is given off the P_{50} will decrease more in the capillaries of mouse lungs than of human lungs. This raises another question: is the Bohr-shift rapid enough so that the exchange process may draw full advantage of it? The rate limiting step of the Bohr-shift is the hydration of CO_2. This reaction depends on the activity of carbonic anhydrase. It may be significant that Larimer and Schmidt-Nielsen (1960) found the carbonic anhydrase activity to increase with decreasing size (Table 5.3).

These thoughts on the respiratory properties of blood in relation to gas exchange in the lung of animals of varying size should be considered as a fragile web of possible connections between phenomena. I hope that the numerous dubious points will stimulate experimental work on these problems.

Another aspect of increasing lung size concerns the pattern of air flow during ventilation. Laminar gas flow becomes unstable and tends to be

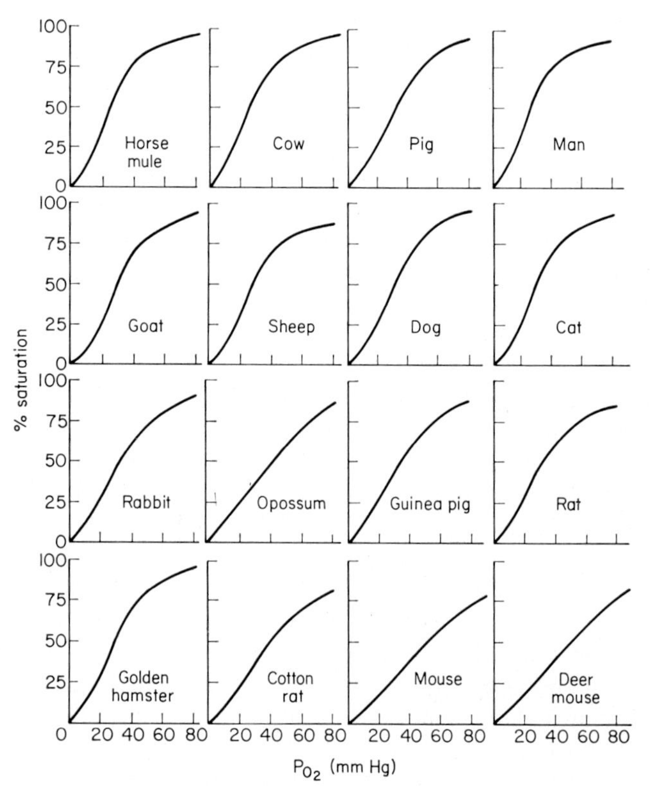

Fig. 5.20. Oxygen-equilibrium curves of whole, unbuffered blood of various mammals (from Schmidt-Nielsen and Larimer, 1958).

turbulent when the product of its velocity and the diameter of the tube reaches a certain value; and the more turbulent the flow the higher is the resistance to its flow. It is likely that one of the factors which limits ventilation during maximal exercise is the transition, especially in the coarser airways, from laminar to turbulent gas flow. We can compare the conditions of gas flow in the trachea of mice and humans. We simplify the problem, possibly too much, by assuming the trachea to be rigid and the flow to be even. The velocity is equal to the flow V, divided by the cross-sectional area A.

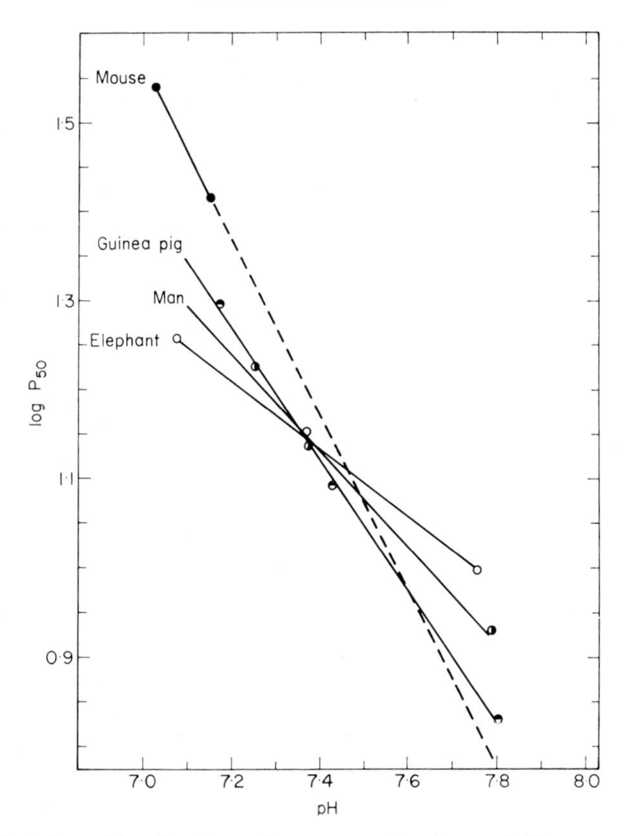

Fig. 5.21. Variation of log P_{50} for various mammalian haemoglobins as a function of pH. Pressure in mm Hg (from Riggs, 1959–60).

$$v = \frac{V}{A}$$

where V is twice the minute volume and A is πr^2, r being the radius of the tracheae.

Using the values from Table 5.4 we get about 165 cm/sec for both tracheae. Since the diameter of the human trachea is about 20 times larger than the mouse trachea, turbulence should therefore occur far more frequently in man

TABLE 5.4
Ventilation and tracheal radius of mice and men.

	Vent ml/mm	Tracheal radius mm
Mouse	23	0·4
Man	7500	7

than in mice. Two factors tend to counteract this. One is the fact that in the mouse lung the points of bronchial branching must lie closer together. Local turbulences and/or eddies will most likely occur at each branching. Furthermore the high ventilation frequency of mice will most likely produce turbulence at lower gas velocities than the low ventilation frequency of man.

We can distinguish two factors which appear particularly relevant in connection with the construction of lungs of varying sizes. One is the efficiency of respiration (i.e. the ratio of total O_2-uptake to the O_2-consumption necessary for ventilation), the other is the "controllability" of ventilation.

The energy of respiration is used to ventilate the lungs; partly to expand the thorax and lung tissue and partly to overcome the resistance to air flow. The latter point is obviously related to the design of the lungs. Resistance to laminar air flow is measured as the ratio of the pressure difference per cm (ΔP) divided by the flow (V) and is related to the length (λ) of the system, and the radius of the tubes (r) in the following way.

$$\frac{\Delta P}{V} = \frac{1}{r} 4_K \text{ when } K \text{ is a constant equal to:}$$

$$\frac{8}{\pi} \times \text{viscosity of the gas.}$$

We shall compare the tracheal resistance in the lungs from mice and man. It is convenient to express this as the pressure fall per cm of trachea necessary to create the normal flow. Relative values for this gradient can be obtained by:

$$P = \frac{V}{r4}$$

$$\text{For the human lung } P = \frac{7500}{0{\cdot}7^4}$$

$$\text{and for the mouse lung } P = \frac{23}{0{\cdot}04^4}$$

The gradient comes out to be 300 times steeper in the mouse trachea than in the human trachea. Despite the numerous assumptions and approximations involved it appears safe to conclude that the pressure gradient is at least one order of magnitude steeper in the mouse trachea than in the human trachea. This possibly applies to the total flow resistance of the two lungs.

In view of the fact that the ventilation path is more than 10 times longer in man than in the mouse, it appears probable that both lungs are ventilated by a negative pressure of the same magnitude. A pulmonary pressure difference of this magnitude may be optimal for the control of breathing. A more open lung might make ventilation vulnerable to external pressure changes, whereas a less open one would imply too high resistance.

The similarity of ventilation pressure in lungs of mice and humans indicates that the relative cost of breathing is the same since the O_2-extraction is the same. This again indicates that their lungs operate at the same efficiency.

Variation in respiratory properties of human blood. Recently Parer (1970) described blood gas values in human subjects with Hb's showing abnormal P_{50}. The P_{50} values ranged from 12 to 70 mm Hg (Fig. 5.22). The important point is that despite this wide difference in P_{50}, cardiac output, venous P_{O_2}, arterial P_{O_2} and arterio-venous difference in O_2-content was much the same. This was achieved by compensation in the O_2-capacity of the blood: Subjects with low P_{50} had high O_2-capacity and *vice versa.* As a result almost complete saturation was found in arterial blood with low P_{50}, but only 60% saturation was found in subjects with P_{50} of 70 mm Hg (Fig. 5.23).

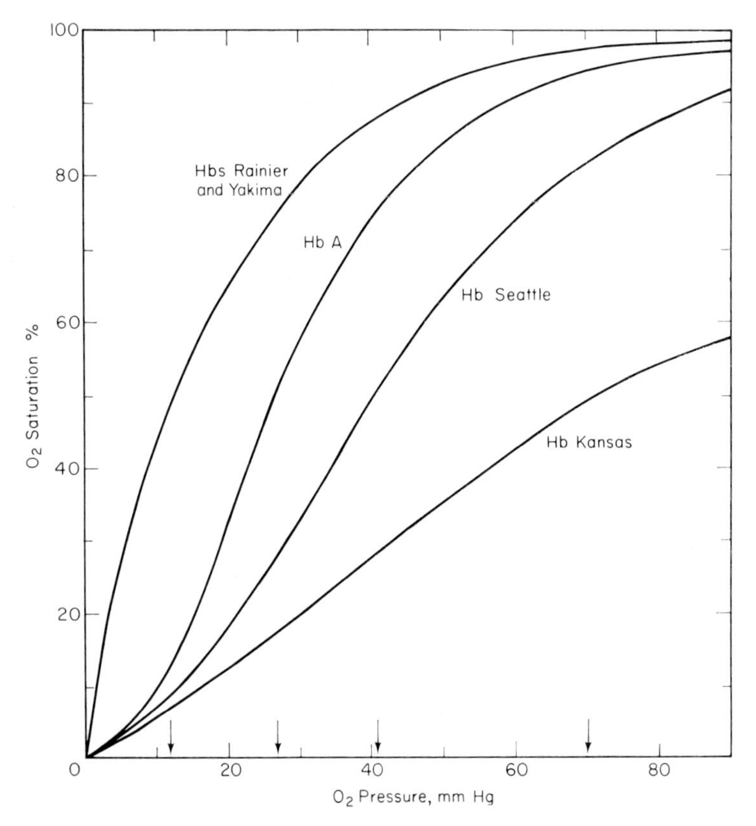

Fig. 5.22. Blood-O_2-dissociation curves of some abnormal haemoglobins. Arrows indicate P_{50}. Notice the inverse relationship between P_{50} and O_2-capacity (from Parer, 1970).

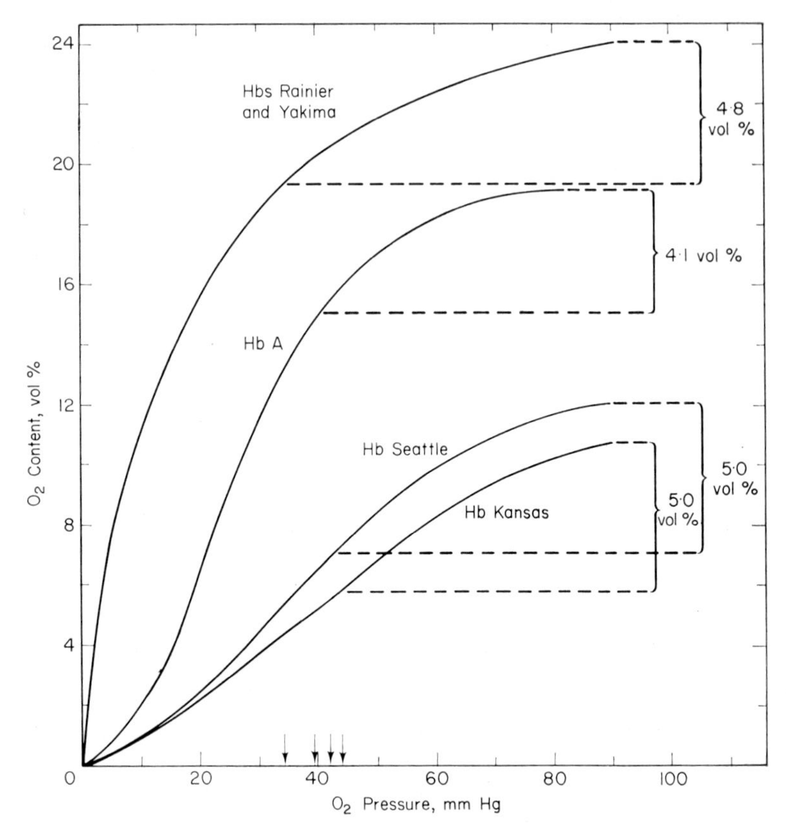

Fig. 5.23 Blood-O_2-dissociation curves of abnormal human haemoglobins. Arrows mark venous P_{O_2}-content difference. See text for details (from Parer, 1970).

E. The Air Capillary Lung

The structural specialization of lungs beyond that found in lungfishes has taken two routes. One is the alveolar type, the other the capillary type. The latter is found in birds. However, in some lizards, turtles and crocodiles the lungs bear a close resemblance to the bird lung. In these the lung has a main central tube which is continuous with the bronchus. From this central tube, branches lead outward in a more or less radial fashion. In bird lungs this basic structure has been further elaborated.

Bird lungs are first of all distinguished from the mammalian lung by the ventilatory pattern. While in the latter air is pumped into and out of the lung, in bird lungs the air is pumped through the lung, into the air sacs and back again. From the point of view of gas exchange, however, the two patterns may not be very different since in both cases the respiratory surface proper is made up of closed structures, alveoli and air capillaries, which differ mainly in form.

1. ORGANIZATION OF THE AIRWAYS

The lungs of a bird occupy a much smaller part of the body volume than the lungs of mammals. Thus a 1-kg crow has a lung volume of 10 ml (Hazelhoff, 1951) while a 1-kg rabbit has a lung volume of at least twice this value.

The structure of the bird lung is well described in textbooks (Sturkie, 1954; Salt and Zeuthen, 1960). A single trachea leads from the mouth where air enters partly through the nostrils and partly through the mouth—which route it takes depends upon the conditions. The trachea branches into the two primary bronchi which enter the lungs, and expand slightly to form a vestibulum. On entering the lungs each of these gives rise to several secondary bronchi (mesobronchi). These give off branches called tertiary bronchi (parabronchi). Neither of these two types of bronchi have typical respiratory

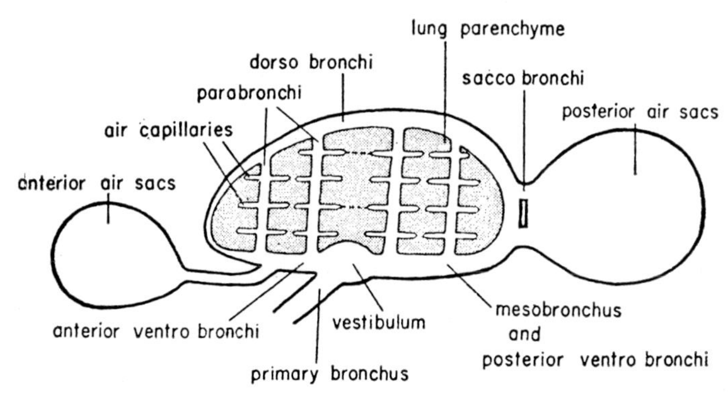

Fig. 5.24. Diagram of the air passages of the respiratory system seen in lateral view (from Salt and Zeuthen, 1960).

epithelium, but are covered by ciliated, simple columnar epithelium. The tertiary bronchi penetrate the lung parenchyma in a dorsoventral direction. The tertiary bronchi are rather straight tubes which tend to anastomose, particularly in the frontal plane of the middle of the lung. Fig. 5.24 is a schematic representation of the bird lung.

Salt and Zeuthen (1960) describe the avian lung in the following way: "Imagine the lung parenchyma held between the hands. The spread fingers represent the ramified network of secondary bronchi on the dorsal and ventral sides of the lung. Each wrist represents either ventral or dorsal bronchi. The tips of the fingers touch slightly along the edges of the lung. The hands are so placed relative to each other that the fingers are not parallel and so that the wrists are not directly above each other. The wrists and the fingers, of course, represent hollow tubes and the two wrists extend from different levels of another tube, the primary bronchus, which connects the tubes and which is itself connected to one end to the upper air tract and at the other end to the large posterior air sacs. Distally the fingers are connected directly to the lung by numerous dorsal-ventral tertiary bronchi." Each tertiary bronchus is the central axis of the respiratory unit of the lung. Along its length it gives off a number of respiratory or air capillaries radiating in all directions. This can be visualized as a test-tube brush, although each hair is also branched. Each air capillary arises as a single tube which branches repeatedly as it proceeds outwards. The terminal end of the capillary is closed and entwined with blood capillaries, where they may anastomose with other tubes from the same tertiary bronchus. In birds that are habitual fliers the air capillaries from one tertiary bronchus anastomose freely with those from adjacent ones with no sharp demarcation of respiratory units. In species that do not fly much, there is little or no such anastomosis, and each unit is sharply limited by a connective-tissue septum.

The air capillaries are lined by simple cuboidal epithelium which becomes progressively thinner as the capillaries branch and decrease in diameter towards the ends, before finally becoming indistinct and fusing with the epithelium of the blood capillaries. Thus the structure of the tips of the air capillaries appears to be almost identical to that of the proper alveoli of mammalian lungs.

In the chicken the diameter of the tertiary bronchi is about 0·2 mm, while that of the air capillaries is about 10–20 μ. The blood to gas distance is less than 1 μ in the capillaries.

Hazelhoff (1951) estimated that the volume of the lungs of crows was about 10 cm^3 with a diffusion area of 200 cm^3. Thus the ratio of volume to area in these units is 20. For comparison the volume of a human lung is 5000 cm^3, while the area is 750 000 cm^2; thus the area to volume ratio is 150.

The abdominal, thoracic and anterior air sacs are connected to the air system of the lung by continuations of the secondary bronchi. The air sacs do not have a blood supply that indicates gas exchange, nor are they covered by thin respiratory epithelium.

2. GAS EXCHANGE WITHIN THE RESPIRATORY UNIT

The respiratory unit of the bird lung is composed of a central tertiary bronchus and its radial, branching and closed air capillaries which penetrate the lung tissue. The resemblance of this system to the tracheal system of insects should be appreciated. The question has been raised whether the air capillaries are ventilated. Zeuthen (1942) and Hazelhoff (1951) investigated the problem by calculating the difference in P_{CO_2} that would have to exist between the ventilated bronchi and the farthest end of the air capillaries to assure sufficient gas renewal at the end of the capillaries. Both authors concluded that diffusion alone was sufficient for ample gas exchange. This also applied during the most severe exercise. However, this does not exclude that other factors may facilitate gas renewal in the air capillaries, but it makes the additional factors less interesting.

Zeuthen's (1942) calculation was based on the steady-state form of Fick's law:

$$V = D\frac{A\Delta P,}{l} \text{ or for this purpose: } \Delta P = \frac{Vl}{AD}$$

He used the following values:

V = CO_2 output = one-half of total CO_2-production = 0·2 ml/sec

l = length of diffusion path = average length of air capillaries = 0·025 cm

A = total cross-sectional area of air capillaries in one lung = 50 cm^2

D = diffusion coefficient for CO_2 in air = 0·15 $\frac{ml}{sec \times cm \times atm}$

ΔP = difference in CO_2-pressure between gas in the tertiary bronchus and in the distal end of air capillaries in atmospheres.

With these values ΔP_{CO_2} comes out to be 0·0007 atmospheres or 0·5 mm Hg. In other words if the ΔP_{CO_2} between the two gases is 0·5 mm Hg then ample CO_2-exchange within the respiratory unit will occur. If the diffusion distance is assumed to be 4 times the assumed value and CO_2-exchange increased 25-fold, which it may be during flight, the ΔP_{CO_2} would equal 0·07 atmosphere or 53 mm Hg. Thus, as was the case with gas exchange in the tracheal system of the *Cossus* larvae (Krogh, 1941), it appears that sufficient gas exchange will take place by diffusion.

3. VENTILATION

A detailed discussion of ventilation in birds is found in Zeuthen (1942). Under normal conditions inspiration and expiration follow one another without intervening pauses. They both last approximately the same length of time. Both inspiration and expiration are brought about by muscle contraction and ventilation is thus entirely active. The elastic forces tend to return the thorax to a rest position.

During inspiration the angle between vertebral and external ribs increases, whereby the sternum is forced forward and downward. This causes expansion of the thorax and air is sucked through the lungs which are fastened to the dorsal ribs. During expiration the contraction of the expiratory muscles reverses this movement and contracts the body cavity.

The organization of the respiratory system of birds implies a different pattern of ventilation from that which is found in mammals. Primarily, this is due to the fact that in birds the lungs are thoroughfare organs between the atmosphere and the system of air sacs. Several techniques have been used to measure the volume of the lungs and of the air sacs in birds. Table 5.5 shows the results. Data on the gas composition of the various sacs are shown in Table 5.6 and the percentage partition of inspired air among the sacs in Table 5.7. Although the applicability of the values to the normal physiological situation may be doubtful, they show quite clearly that the air in the lungs amounts to no more than 10–15% of the total volume of the entire system.

The fact that the air sacs always contain some gas is most likely an adaptation to provide the bird with a desirable specific gravity. From a respiratory point of view it is particularly important when considered in connection with the distribution of inspired air between the respiratory and the thoroughfare paths of the lung and among the air sacs.

TABLE 5.5
Volumes in ml of the air sacs of some birds (from Salt & Zeuthen, 1960).

Species	Inter-clavicular	Pre-thoracic Left	Pre-thoracic Right	Post-thoracic Left	Post-thoracic Right	Abdominal Left	Abdominal Right	Lung	Dead space
Chicken	9	9	8	3·5	3	25	37	3	1·6
Duck	53	11	13	27	30	65	80	—	4
Pigeon	—	2·8	3·1	1·4	2·2	10·1	8·6	—	—

4. DISTRIBUTION OF GAS

Gas distribution in the bird lung has long been a matter of dispute. The reason for this seems to be the difficulty with which experimental data can be obtained. The pattern of distribution seems to be very labile and dependent upon the condition of the bird more than is the case in mammals. Earlier many investigators favoured the view that the distribution of gas in the bird lung depended on particular gas valves. However, no such anatomical structures have been found and today the accepted view is that the distribution of gas in the bird lung occurs in principally the same way as blood distribution, namely by variation of the diameter of the tubes in different sections of the lung. Salt and Zeuthen (1960) give an admirably clear account of the different viewpoints that have been, or still are, in vogue.

The distribution of the gas in the lungs during inspiration and expiration has important functional implications. If the entire inspired volume ventilates the respiratory units during both phases, then the blood which circulates the lung during inspiration will equilibrate at higher P_{O_2} and lower P_{CO_2} than the blood circulating during expiration. The resultant rhythmic variation in blood composition could be avoided by compensatory adjustments of blood flow. The design of the airways with an abundant possibility to shunt gas around the respiratory units suggests, however, that only a proportion of

TABLE 5.6

The percentage composition of gas in the air sacs of some birds (from Salt and Zeuthen, 1960).

Species	Cervical sac CO_2%	Cervical sac O_2%	Interclavicular sac CO_2%	Interclavicular sac O_2%	Pre-thoracic sac CO_2%	Pre-thoracic sac O_2%	Post-thoracic sac CO_2%	Post-thoracic sac O_2%	Abdominal sac CO_2%	Abdominal sac O_2%	Expired CO_2%	Expired O_2%
Pigeon			4·5	15·5	4·8	15·0	4·7	15·8	3·9	16·5	3·82	14·71
Duck			6·4	12·1					2·2	18·5	4·8	13·7
Chicken	3·2	15·6	5·0	14·6	3·2	16·3	3·4	17·4	2·0	19·0		

TABLE 5.7

Percentage of inspired air that goes to the different air sacs (from Salt and Zeuthen, 1960).

Species	Cervical	Inter-clavicular	Pre-thoracic Left	Pre-thoracic Right	Post-thoracic Left	Post-thoracic Right	Abdominal Left	Abdominal Right	Total accounted for
Chicken	—	—	—	6%	6%	2%	33%	47%	94%
Duck	—	1%		3%	23%		49%		76%

the gas ventilates the exchange areas. *A priori* two patterns of distribution appear to give the desired constant blood composition. One is that half the inspired volume goes directly to certain air sacs, while the other half goes through the respiratory tissue and into other sacs. Upon expiration the sacs containing air supply the respiratory tissue, while the others empty directly through the trachea. Such a situation requires that some air sacs contain a mixture of dead space gas and air, others expiratory gas. Direct measurements show that this is not so. The composition of gas in different sacs is remarkably uniform (Table 5.6).

TABLE 5.8

The calculated percentage of total flow through tertiary bronchi assuming different values for CO^2-concentration in gas leaving tertiary bronchi. Numbers in parentheses less dependable (from Zeuthen, 1942).

Subject	% CO_2 in gas leaving tertiary bronchi	% of total gas flow through tertiary bronchi during:	
		Inspiration	*Expiration*
Hen 1.2	6	45	(89)
	7	38	(64)
	8	32	52
	9	29	44
		(26)	38
Hen 3	6	48	(126)
	7	41	(87)
	8	36	67
	9	31	54
	10	(31)	46

To understand the significance of the other, and experimentally the most substantiated possibility, we must appreciate that gas passes through a tissue of considerable length. While doing so the P_{CO_2} will increase and the P_{O_2} decrease. The arterial gas composition will be related to the average composition of gas passing through the lungs. Since the venous P_{CO_2} is constant the change in tension along the respiratory unit will be steeper the lower the initial gas P_{CO_2}. Also the rate of change in gas composition, which determines the average tension, will depend on the rate of gas flow. Zeuthen (1942) analyzed such a model quantitatively. His main problem was: what percentage of the total inspired and expired gas passes the tertiary bronchi? He calculated this based on experimental values for P_{CO_2} of arterial blood, P_{CO_2} in the air sacs, volume of air sacs, volume of dead space and of air passages in the lungs, the tidal volume and the P_{CO_2} of expired air.

Table 5.8 shows the calculated percentages of the total ventilated volume which pass the tertiary bronchi during inspiration and expiration. During inspiration between 30 and 50% of the total volume ventilates tertiary bronchi, during expiration from 40 to 70%. Thus more of the gas passes tertiary bronchi during expiration than during inspiration.

Gas distribution in the respiratory system of birds has quite recently been reinvestigated. The results do not confirm Zeuther's conclusion. Schmidt-Nielsen *et al.* (1969) investigated the problem in the ostrich. Its size allows measurements which are very difficult in smaller animals. They found, contrary to earlier information, that the composition of gas in the different air sacs is not the same. The anterior sacs contained typical expiratory gas (6–8% CO_2, 13–15% O_2) while the posterior sacs were composed of 3–4% CO_2 and 17% O_2. When the ostrich was allowed a single breath of pure O_2, the P_{O_2} in the posterior sacs increased at the end of the breathing cycle, whereas the P_{O_2} of the anterior sacs increased less and only 1–3 cycles later. They suggested that during inspiration gas passes through the lungs into posterior sacs without passing, to a significant extent, gas exchange surfaces. From these the gas passes respiratory lung epithelium and enters the anterior sacs. Upon expiration gas from the anterior sacs enters the trachea without further exposure to exchange surface. At the same time gas from the posterior sacs passes to the anterior sacs via gas exchange surfaces. This pattern of flow and distribution has also been found in ducks by using directional micro-thermistor probes *in situ*. The same pattern was found whether the bird was fully awake at normal temperature, panting or anaesthetized (Bretz and Schmidt-Nielsen, 1970). The authors point out that this pattern allows the flow of gas through the gas exchange units to be the same during inspiration and expiration. Counter-current exchange appears nevertheless to be unimportant since the gas at the exchange surfaces is as stagnant as in the alveoli.

Recently quantitative information concerning respiration during flight has been obtained thanks to the development of telemetric techniques (Fig. 5.25). Thus Hart and Roy (1966) were able to measure frequency and depth of ventilation as well as heart rate and wing beat in flying pigeons. Their data largely support the predictions made by Zeuthen (1942) from his work on resting birds. Berger *et al.* (1970) investigated several other species. In all of them ventilation was coordinated with the wing beats, but not always in a 1 : 1 coordination. Birds with small wings, for example quail, had 5 wing beats per ventilation. In all species the peak of expiration was found to coincide with the bottom of the downstroke. Compared to the resting state, flight is accompanied by about 8-fold increase in heat production but 20-fold increase in pulmonary ventilation (Fig. 5.26, Table 5.9). The increase in ventilation is due exclusively to an increase in frequency. Upon landing ventilation frequency drops but the tidal volume, that is the volume of each breath, increases. As a result the minute volume decreases only slowly.

In birds, as in other homeothermic animals, ventilation may serve to regulate temperature as well as exchange gases. In birds this dual purpose is more strikingly shown than in other animals. Water loss from the lungs has been measured directly during flight by LeFebvre (1964) by the use of H_2O^{18}. Hart and Roy calculated water loss on the assumption that inspired air at 25°C was humidified and expired air at a temperature of 40°C was saturated with water (Table 5.9). Also Pearson (1964) calculated water loss

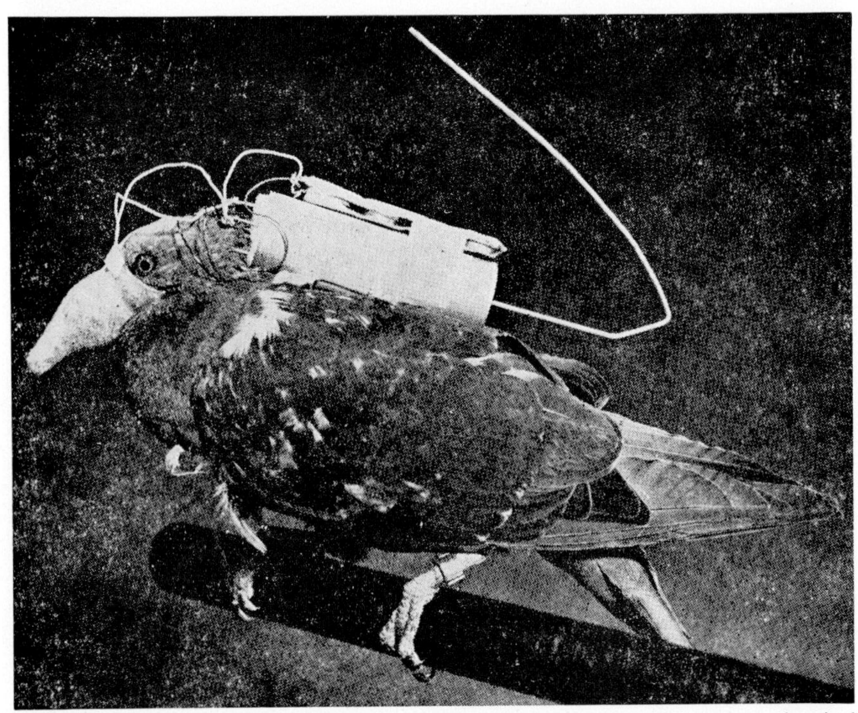

Fig. 5.25 A pigeon equipped with mask and electronic transmitter which conveys electrical signals from implanted electrodes, while the bird is flying, to a receiver on the ground (from Hart and Roy, 1966).

during flight from weight-loss measurements. All three methods have given values in the same range: 10 g per kg body weight per hour. Hart and Roy (1966) found that during flight the evaporative heat loss was responsible for about 17% of total heat loss, while during rest it was responsible for about 7%. The increased portion of the total heat production given off by evaporation from the respiratory surfaces is in accord with the increase in ventilation. While the heat production increases only 8 times, ventilation increases 20 times. Most likely this extra ventilation does not affect the tertiary bronchi but goes directly through the lungs to air sacs where it is humidified and returns via the same route. Unfortunately values are not yet available on the

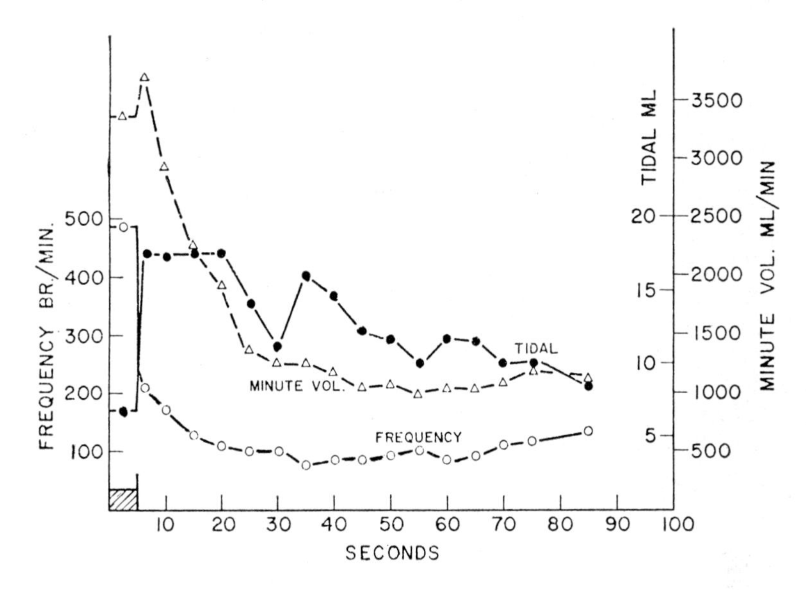

Fig. 5.26. Changes in breathing frequency, tidal volume and minute volume recorded for an 80-sec postflight period. Flight period is indicated by *crosshatched mark* at 0–5 sec. Recovery is characterized by a rapid fall in frequency, a rise in tidal volume followed by a slow decline, and a gradual fall in minute volume (from Hart and Roy, 1966).

TABLE 5.9

Heat production, ventilation and respiratory evaporation during rest and flight in pigeons (from Hart and Roy, 1966).

				Evaporative heat loss		
	Heat production (cal/h) (1)	Pulmonary ventilation (litres/h) (2)	H_2O^b loss (g/h) (3)	cal/h (4)	% Heat production (5)	% Estimated expired CO_2 (6)
Rest	2.69	7·2	0·31	0·18	6·7	7·5
Flight	22·0	147·0	6·34	3·68	16·7	3·0
Flight rest	8·2	20·4	20·4	20·4	2·5	—

blood composition during flight, but we must assume that in parallel to what is the case in other homeotherms, birds will also, during flight, maintain a constant arterial P_{O_2} and P_{CO_2}.

Chapter 6

RESPIRATORY ADAPTATIONS TO LIFE AT HIGH ALTITUDE

Of the naturally occurring hypoxic conditions, life at high altitude is the most permanent. Certain species have lived in such environments for thousands of years. Their respiratory mechanisms should therefore be well adapted to the situation. It is regrettable that only a few students of respiratory physiology have investigated the respiratory physiology of high-altitude animals other than man. This species has, on the other hand, showed great talents in combining research with exciting expeditions to remote areas. His pleasant efforts have provided a good description of the main reactions and adaptations to altitude (Kellogg, 1968).

The main environmental stress factor of high altitude, which is of interest here, is the low partial pressure of O_2. A very popular location for high-altitude work is the Peruvian mining town of Morococha, elevation 4530 m, P_{O_2} of about 85 mm Hg. This ambient P_{O_2} is thus below the alveolar P_{O_2} of "lowlanders".

Hurtado (1964) reviews several investigations which show that O_2 uptake both at rest and during work is the same in Morococha as it is in Lima which lies at sea level. Thus the problem is to explain how the decreased P_{O_2}-difference between ambient air and tissue at high altitude is compensated for by other respiratory parameters to achieve the same O_2-uptake. The total P_{O_2}-difference from inspired air to venous blood can be divided into steps as shown in Fig. 6.1. We should be aware that the quantitative determination of P_{O_2} at all points except ambient air, arterial blood and mixed venous blood is so difficult that the values are uncertain and may introduce errors in this analysis. We must, however, discuss the subject as if the values are correct.

The P_{O_2} decrease from air to alveolar gas is smaller in individuals breathing in Morococha than in Lima. This is due to at least two factors. One is a 20–40% increased ventilation, the other an increased ratio of alveolar volume to total lung volume (Hurtado, 1964).

The ΔP_{O_2} from alveolar gas to venous blood is about 5 times greater in lungs ventilated in Lima than in Morococha. The ΔP_{O_2} between alveolar gas and mean capillary P_{O_2} is 4 times smaller in Morococha than in Lima. Thus at a 4 times smaller ΔP_{O_2} the O_2 uptake is still the same. The reason why ΔP_{O_2} is larger between alveoli and venous blood than between alveoli and mean capillary blood, is that at high altitude a steeper part of the O_2-dissociation curve is in use.

This apparent discrepancy between ΔP_{O_2} and O_2-uptake at high and low altitude is explainable in a semi-quantitative way by a combination of several co-operating factors, all of which ultimately must contribute to increase the diffusion area and/or reduce the distance.

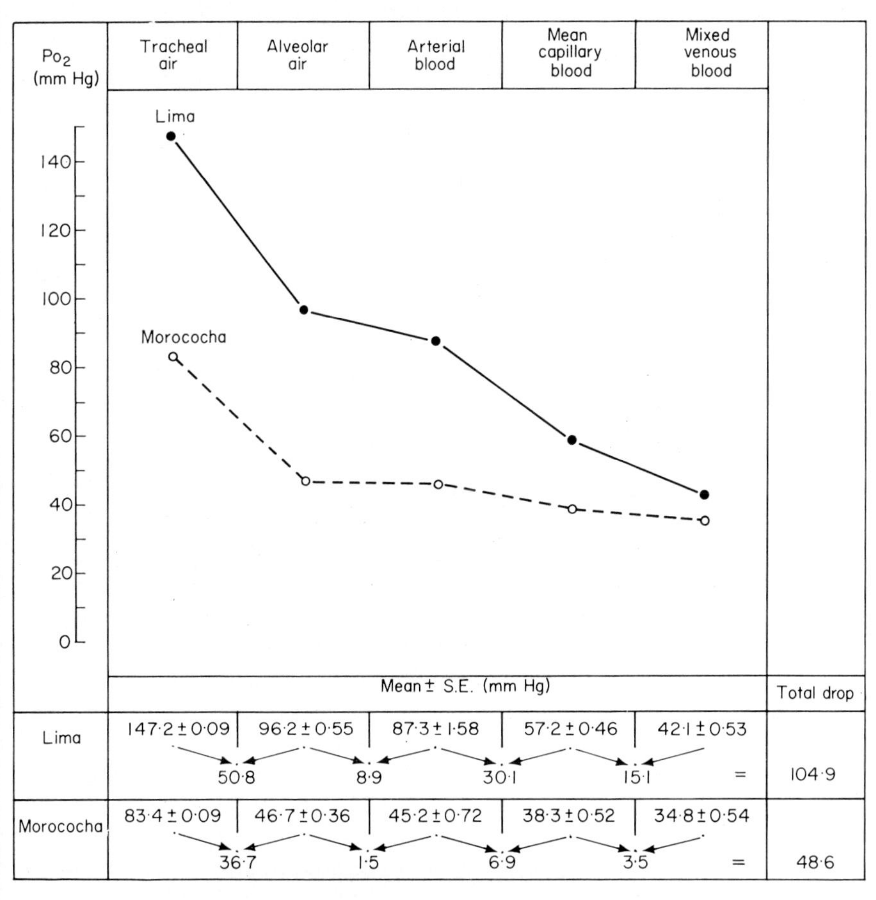

	Tracheal air	Alveolar air	Arterial blood	Mean capillary blood	Mixed venous blood	Total drop
Lima	147·2 ± 0·09	96·2 ± 0·55	87·3 ± 1·58	57·2 ± 0·46	42·1 ± 0·53	
	50·8	8·9	30·1	15·1	=	104·9
Morococha	83·4 ± 0·09	46·7 ± 0·36	45·2 ± 0·72	38·3 ± 0·52	34·8 ± 0·54	
	36·7	1·5	6·9	3·5	=	48·6

Fig. 6.1. Mean P_{O_2} pressure gradients, from tracheal air to mixed venous blood, in native residents of Lima (sea level) and Morococha (4540 m). Mean values correspond to two groups of eight healthy adult subjects each, studied at sea level and at high altitudes. Alveolar air, arterial blood, and mixed venous blood (from the pulmonary artery) were obtained simultaneously, at rest in the recumbent position. Mean capillary blood P_{O_2} was calculated (from Hurtado, 1964).

The previously mentioned increase in total alveolar volume probably involves both recruitment of new alveoli as well as increased diameter of each alveolus. If we oversimplify the alveolar-capillary barrier and consider it a homogeneous half-sphere we can calculate that if it is stretched so that the diameter increases from 200 to 280 μ, the wall thickness will decrease to one-half its previous value. This will also double the alveolar area and increase the volume by a factor of about 3. Whether an increase in alveolar diameter of this magnitude really takes place, is not known. Neither do we know much about the distribution of gas in the lungs at high altitudes.

To achieve similar O_2-delivery in the tissues at high and at low altitude in face of a smaller gradient also requires an increased ratio of area to diffusion distance. This is achieved at least qualitatively by increased capillarization in the muscles (Valdivia, 1956).

TABLE 6.1

Some pertinent respiratory parameters of persons living at high and low altitude (from Hurtado, 1964).

	Lima (sea level)	Morococha (4540 m)
Blood volume ml/kg body weight	80	100
Plasma volume ml/kg body weight	42	39
Haematocrit percentage	47	60
Hb concentration g/100 ml	15·6	20·1
O_2-capacity mm/litre	9·3	12·3
O_2-saturated arterial blood %	98	81
P_{CO_2} arterial blood	40	33
pH plasma	7·41	7·39

Adaptation to high altitude also causes changes in two of the basic respiratory properties of blood: the O_2-capacity and O_2-affinity. It has been shown (Hurtado, 1964) that the O_2-capacity of the blood of permanent residents of Morococha is about 25–40% higher than that of sea-level residents (Table 6.1). The increased O_2-capacity is due to increased rate of red cell production and causes the haematocrit to reach about 60%. In some permanent residents of Morococha, values up to 85% have been recorded. Such values are abnormal and subjects having them suffer from various complications. Burton and Smith (1969) found increased red cell concentration in chicks developing at high altitude.

Fig. 6.2 shows the O_2-equilibrium curves of blood from lowlanders and high-altitude residents. Normal values for arterial and venous P_{O_2} at the two altitudes are shown. We notice that high-altitude adaptation results in increased O_2-capacity, but decreased O_2-affinity. A similar reaction is found in some types of anaemia (Mulhausen *et al.*, 1967).

In order to appreciate the "respiratory significance" of these two changes we shall consider the following hypothetical situation:

A lowlander arrives to Morococha. The first hour the respiratory properties of his blood are unchanged (P_{50} = 27 mm Hg, O_2-capacity 20 volume per cent) then suddenly the O_2-affinity of his blood is decreased so that his P_{50} is 30 mm Hg. After another hour his O_2-capacity is raised from 20 to 27 volume per cent, while the P_{50} drops back to 27 mm Hg. After another hour his P_{50} increases to 30 mm Hg and his capacity is maintained at 27 volume per cent. What effect do these changes have on the ΔP_{O_2} from alveoli to capillary blood and from capillary blood to tissue?

We shall make the following simplifying assumption: The alveolar and arterial P_{O_2} remains constant at 46 mm Hg and 45 mm Hg respectively, the A–V difference in blood O_2-content remains constant at 4 volume per cent and the "tissue P_{O_2}" is constant at p.

Table 6.2 is based on these assumptions and the data presented in Fig. 6.1 and Fig. 6.2. The mean ΔP_{O_2} from alveolar gas to capillary blood is 17 mm Hg in the newcomer at Morococha, drops 4 mm Hg owing to increase in P_{50}

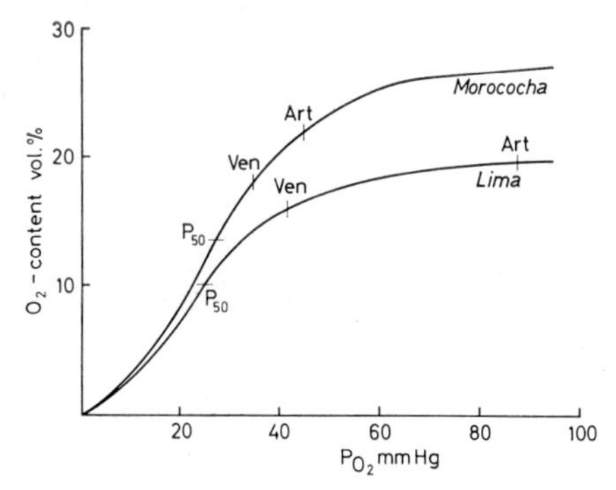

Fig. 6.2. Oxygen-equilibrium curves for blood from persons living at high altitude, Morococha, and at sea level, Lima. Normal values for arterial and venous blood are shown as are the P_{50} (modified from data of Hurtado, 1964).

amd 3 mm Hg owing to increase in O_2-capacity. Together these two changes will decrease the pulmonary gradient by 7 mm Hg or by about 40%. The blood to tissue ΔP_{O_2} will increase by 4 mm Hg due to increased P_{50} and by 3 mm Hg owing to increased O_2-capacity. Together these factors increase the blood to tissue gradient by 7 mm Hg. The relative increase will depend on the tissue PO_2. If this is 10 mm Hg then the increased P_{50} will cause an increase in the gradient of 20%, while the increase in O_2-capacity will increase the gradient by 15%. It is evident that the adaptive changes in O_2-affinity and O_2-capacity are about equally important. Since these changes tend to decrease the pulmonary ΔP_{O_2}, but increase the tissue ΔP_{O_2}, it seems that gas exchange in the tissues must be the limiting step in the overall gas exchange process.

The O_2-affinity of blood depends on two factors: the nature of the Hb-molecule and the intracellular environment surrounding the molecule. There

TABLE 6.2

Shows the effect of adaptive changes to high altitude in O_2-capacity and O_2-affinity on the partial O_2-pressure across the pulmonary surface and in the tissue. For details see text.

Blood properties		Arterial blood		Venous blood		Capillary blood	Alveolar	Tissue	P_{O_2}	
P_{50} mm Hg	O_2-cap. vol. %	P_{O_2}	O_2-content	P_{O_2}	O_2-content	Mean P_{O_2}	P_{O_2}	P_{O_2}	alveoli/ capacity	capacity tissue
27	20	45	16·4	29	12·4	37	46	p	17	29–p
30	20	45	16·4	33	12·4	39	46	p	13	33–p
27	27	45	22·2	32	18·2	38	46	p	24	32–p
30	27	45	22·2	36	18·2	40	46	p	10	36–p

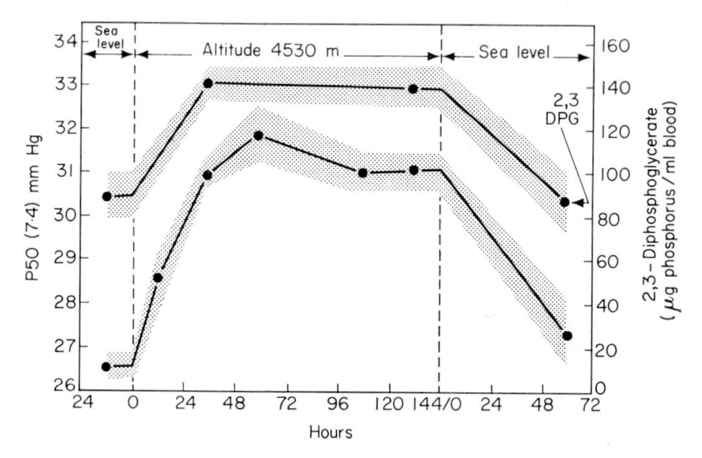

Fig. 6.3. Changes in P_{50} at (pH = 7·4) and 2,3-DPG of whole blood induced by high altitude. Mean values are shown by the lines, one standard deviation is indicated by the shaded areas. Parallel increases with exposure to high altitude and decreases on return to sea level were observed (from Lenfant *et al.*, 1968).

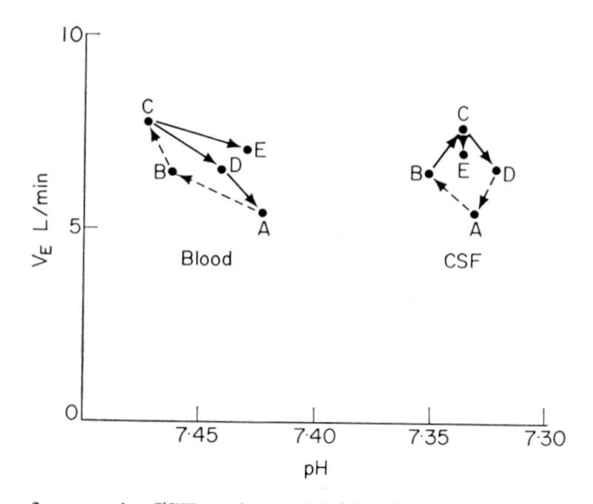

Fig. 6.4. Sequence of events in CSF and arterial blood in ascent to and descent from high altitude. Solid line represents an acid shift in pH which should stimulate the appropriate receptors, dashed line an alkaline shift which should depress them (from Mitchell, 1966).

- AB Effect of acute hypoxia.
- BC Adaptation to hypoxia over the next few days.
- CE Adaptation occurring with prolonged residence at altitude, presumably due to loss of sensitivity of the peripheral chemoreceptors.
- CD Acute return to sea level.
- DA Gradual return of ventilation to normal over the next few days to weeks at sea level.

is little evidence that high-altitude exposure changes the Hb molecule, but recent evidence indicates that it influences the internal chemical environment of the red cell. Benesch *et al.* (1968) found that 2,3-diphosphoglycerate (2,3-DPG) influenced the O_2-affinity of Hb in solution. Lenfant *et al.* (1969) showed that the increase in P_{50} during the first days of high-altitude exposure was paralleled by an increase in intracellular concentration of 2,3-DPG (Fig. 6.3). The content of 2,3-DPG in permanent high-altitude residents is also increased, but decreases within hours after descent to sea level as does the P_{50}. No information is available on 2,3-DPG in other animals from such altitudes. It will be interesting to follow future research on naturally occurring substances which modify O_2-affinity in response to other conditions.

Adaptation to life at high altitudes also involves changes in the chemical components of blood which act as stimuli for ventilation. In fact, these changes elucidate the multicomponent nature of the chemical regulation of ventilation. Mitchell (1966) shows an illustrative figure of these events (Fig. 6.4). The immediate response to hypoxia is hyperventilation. This causes decreased arterial P_{CO_2} and an alkaline shift in blood and cerebrospinal fluid. This alkaline shift depresses ventilation. During the first days at high altitude the blood and cerebrospinal fluid pH will return to normal owing to active extrusion of HCO_3^- via the kidneys and from the cerebrospinal fluid. Thus a normal pH is re-established at a lower P_{CO_2}, while ventilation is maintained at 20–40% more than in sea-level residents.

Prolonged exposure to high altitude causes a desensitization towards hypoxia. It appears that if a person has lived the first years of his life at high altitude, he loses the normal response to hypoxia, even if he lives at sea level for years thereafter (Sørensen and Severinghaus, 1968). An explanation of this reaction must be speculative, but it appears reasonable that the same relationship between P_{O_2} and pH sensitivity cannot give adequate stimuli at both low and high altitude since the arterial pH is the same in both cases, while P_{O_2} is very different.

Hypoxia normally causes pulmonary vasoconstriction and high-altitude residents have chronic pulmonary hypertension. Grover (1965) made the interesting observation that cattle of highland stock showed this response to hypoxia to a smaller extent than cattle of lowland stock. This difference was evidently genetically determined, probably through selection for generations.

Comparative investigations on animals from low and high altitudes are scarce. Tenney and Remmers (1963) could not demonstrate any difference in the morphology of lungs from sheep and guinea-pigs living at different altitudes. The alveolar diameter and the ratio of alveolar surface to lung volume were independent of altitude. Hall *et al.* (1936) compared the blood of some high-altitude birds and mammals with that from relatives at sea level. High-altitude species showed a higher blood O_2-capacity and a higher O_2-affinity than related species living at sea level. As to the latter point, this is in contrast to the adaptive changes of humans. It may be that through generations at high altitudes other respiratory adaptations, for example tissue tolerance to hypoxia, may have occurred so that a high O_2-affinity has become the most favourable condition. The experimental techniques of Hall *et al.* (1936) are insufficiently described and a reinvestigation of the problem is necessary.

Chapter 7

SECONDARY ADAPTATIONS OF LUNG BREATHERS TO AQUATIC LIFE

The respiratory problems connected with submersion have been met in two ways. One is by having large O_2-stores and to economize with them during the dive. The other is by supplementing the O_2-stores by accessory gas exchange with water. For obvious reasons, homeotherms make use of only the first possibility while some poikilotherms make use of both.

A. Accessory Water Breathing in Vertebrates

Certain species of turtles supplement their O_2-stores by exchanging gas directly with water. Belkin (1968) measured that an aquatic turtle covered 30% of its O_2-consumption during a dive by ventilation of the bucco-pharyngeal areas. The remaining 70% was covered by gas exchange across the skin. However, the total amount of O_2 thus secured by the turtles submerged in air-equilibrated water did not exceed 10% of the O_2-uptake in air.

During the 1967 Alpha Helix Amazon Expedition Peterson and Bellamy (personal communication) obtained evidence that aquatic turtles (*Podocnemys* sp.) covered about 90% of their O_2-uptake by cloacal respiration when the P_{O_2} of the gas phase above the water fell below 10 mm Hg. This turtle possesses well-developed cloacal bursae which are rhythmically ventilated with water. Other turtles which have similar structures evidently do not use them for gas exchange since they are not ventilated during submersion.

B. Adaptations to Prolonged Submersion

With one exception the physiological adjustments displayed by diving animals are not unique adaptations to their particular behaviour, but rather a perfection of the general defence mechanism to asphyxia found in all vertebrates. The one exception concerns the defence against decompression sickness.

The historical development of this field, which is well described in a review article by Andersen (1966) is most interesting. From 1935 to 1941 two physiologists, Irving and Scholander, ably delineated the pertinent problems, did the crucial experiments and foresaw the answers to almost all the essential questions (see: Irving, 1939; Scholander, 1940).

1. BLOOD PROPERTIES AND O_2-EXCHANGE DURING SUBMERSION

The main O_2-stores of the body are the blood, the muscle pigments and the lungs. Diving animals have, relative to their body weight, about twice the blood volume of non-divers. The O_2-capacity of their blood is generally also higher, although not without exception (Table 7.1). Robinson (1939) found

TABLE 7.1

Blood volume and blood O_2-capacity of some diving and non-diving species (from Andersen, 1966).

Animal	Blood volume % body weight	O_2-capacity vol. %
Pigeon	7·0	21·2
Hen	3·9	11·2
Duck, domestic	10·0	16·9
Guillemot, *Uria troile*	12·3–13·7	26·0
Puffin, *Mormon fratercula*	11·3–12·0	24·0
Penguin, *Pygoscelis papua*	9·0	20·0
Man	6·2– 7·0	20·0
Dog	6·2–10·5	21·8
Horse	7·0–10·7	16·7, 14·0
Rabbit	6·5	15·6
Beaver, *Gastor canadensis*		17·7
Muskrat, *Ondatra zibethica*	10	25·0
Seal, *Phoca vitulina*	15·9	29·3
Sea lion, *Eumetopias stelleri*		19·8
Porpoise, *Phocaena communis*	15·0	20·5
Blue whale, *Balaenoptera musculus**		14·1
Fin whale, *Balaenoptera plupalis**		14·1
Sperm whale, *Physeter catodon**		29·1

* Blood samples drawn from carcasses up to several hours after death.

that O_2 bound to the myoglobin of seal muscles might contribute 47% of the total O_2-store. His conclusion that myoglobin therefore is the most important O_2-store during prolonged diving may be questioned since the O_2-affinity of myoglobin is so high (P_{50} = 3–4 mm Hg). Thus, only if the tissue P_{O_2} falls below these low values will the O_2-myoglobin stores be of considerable value to vital organs. Unfortunately, *in vivo* values of tissue P_{O_2} and pH and *in vitro* values of the P_{50} at these pH values are lacking. We know, however, that myoglobin shows only a slight Bohr-effect. The high myoglobin concentration of divers would indicate a storage function. I would not be surprised if future work were to reveal a tissue P_{O_2} in a range where these stores may be of considerable importance.

The lung volume of divers is not significantly different from non-divers. In fact deep-diving species have rather small lungs. This may be favourable for several reasons. A large lung volume would constitute a cumbersome float during the dive. It is significant that divers have a large tidal volume (volume of one breath) relative to the lung volume. This enables quick renewal of gas between dives. During regular air breathing, divers usually show a slow rate of breathing and a high O_2-extraction (Andersen, 1966).

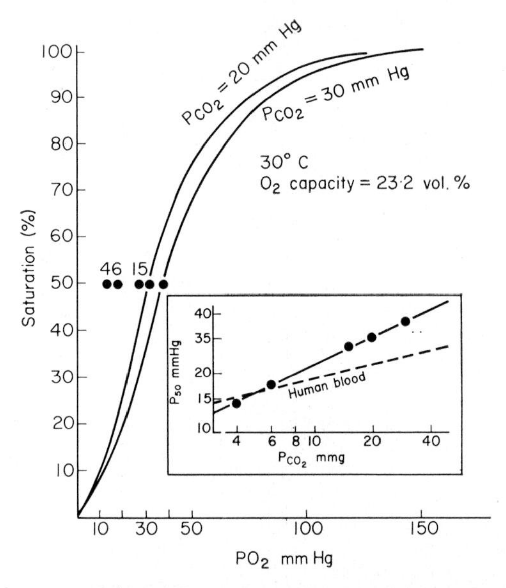

Fig. 7.1. Typical features of blood from divers illustrated by O_2-equilibrium curves of platypus. Insert shows the magnitude of the Bohr-shift compared to human blood. The points to the right of 50% HbO_2 shows P_{50} at the P_{CO_2} given by the numbers above each point (from Johansen *et al.*, 1966).

The O_2-affinity of blood from diving mammals is not typically different from that of non-diving aquatic vertebrates (Lenfant, 1969). The aquatic diving turtle mata mata (*Chelys fimbriata*), however, showed a higher P_{50} than terrestrial turtles (Lenfant *et al.* 1970). Blood from most divers is distinguished by a more pronounced Bohr-effect than non-divers (Table 5.3). Fig. 7.1 shows a typical O_2-equilibrium curve for blood of divers. The pH-sensitive O_2-affinity facilitates O_2-delivery to the tissues, particularly during the later stages of submersion when the acidosis is most pronounced. Andersen and Løvø (1967) showed that at the end of a dive the arterial O_2-saturation in ducks was less than 4% HbO_2. Still, and this is very significant, the P_{O_2} is 25–30 mm Hg, or close to the P_{50} at a normal pH of 7·4. The maintenance of this high

P_{O_2} is due to the Bohr-effect which therefore both causes a large blood tissue ΔP_{O_2} and enables the animal to utilize almost all the O_2 carried by the blood.

Lenfant *et al.* (1970) investigated the respiration of a fresh-water turtle, the mata mata (*Chelys fimbriata*). This turtle stays permanently in water and visits the surface to breath for a couple of minutes at intervals from 20 to 65 minutes. Compared to terrestrial turtles it has blood with low O_2-capacity (about 10 volume per cent as against 15 volume per cent), low O_2-affinity (P_{50} of 26 mm Hg as against 15 mm Hg), and a larger Bohr-effect and buffer capacity than terrestrial turtles. It is also distinguished by large lungs; they comprise 15% of the volume of wet tissue. This has obvious significance as O_2-stores but may also be essential in buoyancy control. At the end of a 30-minute dive the arterial P_{O_2} was still above the P_{50} value but the arterial pH had not decreased by more than one-tenth. This signals an almost complete respiratory adaptation to submersion hypoxia.

2. DEFENCE AGAINST DECOMPRESSION SICKNESS

It is frequently observed that in preparation for a dive seals expire rather than inspire. This appears to be a preparatory defence against decompression sickness (formation of gas bubbles in blood and tissues during decompression). During a dive the lungs of whales are compressed, whereby the lung endothelium shrinks and thickens and supersaturation of the blood is slowed. At 100 m depth the lung volume is at most one-tenth of what it was at the surface. This means that the entire lung volume is contained in the anatomical dead space. A small initial lung volume contained in a lung which is more compressible than the dead space would therefore reduce diffusion of most gas into the blood provided the descent is rapid.

3. CARDIOVASCULAR ADAPTATIONS

The cardiovascular reactions during diving are characterized by reduced heart rate (bradycardia) and redistribution of blood. The degree of bradycardia varies among species and a reduction to one-tenth of normal heart rate is common (Andersen, 1966). The record apparently is held by the lizard *Iguana iguana* whose heart rate at the onset of dive diminishes to less than 1% of normal (Belkin, 1963). It is quite typical that this cardiac retardation is not accompanied by seriously reduced blood pressure (Irving *et al.*, 1942). This, of course, means that the resistance to blood flow is increased in proportion to the decreased cardiac output.

Johansen (1964) measured the relative blood perfusion to various organs. Using radioactive rubidium (Rb^{86}) as an indicator he measured the blood flow to different organs in ducks before and during a dive. He found that the large skeletal muscles such as pectorals and gastrocnemius received an insignificant blood supply during the dive. On the other hand supernormal values were found in the heart and the brain. High values were also found in the eyes and in the tongue and in muscles of the head which are used to catch food during natural dives (Fig. 7.2).

4. ENERGY PRODUCTION

It appears that during a dive mainly the brain and the heart receive an ample blood supply. These organs therefore can obtain energy by aerobic metabolism. The others rely upon anaerobic processes. This is not to say that the circulated regions do not employ anaerobic metabolism. Fig. 7.3 illustrates

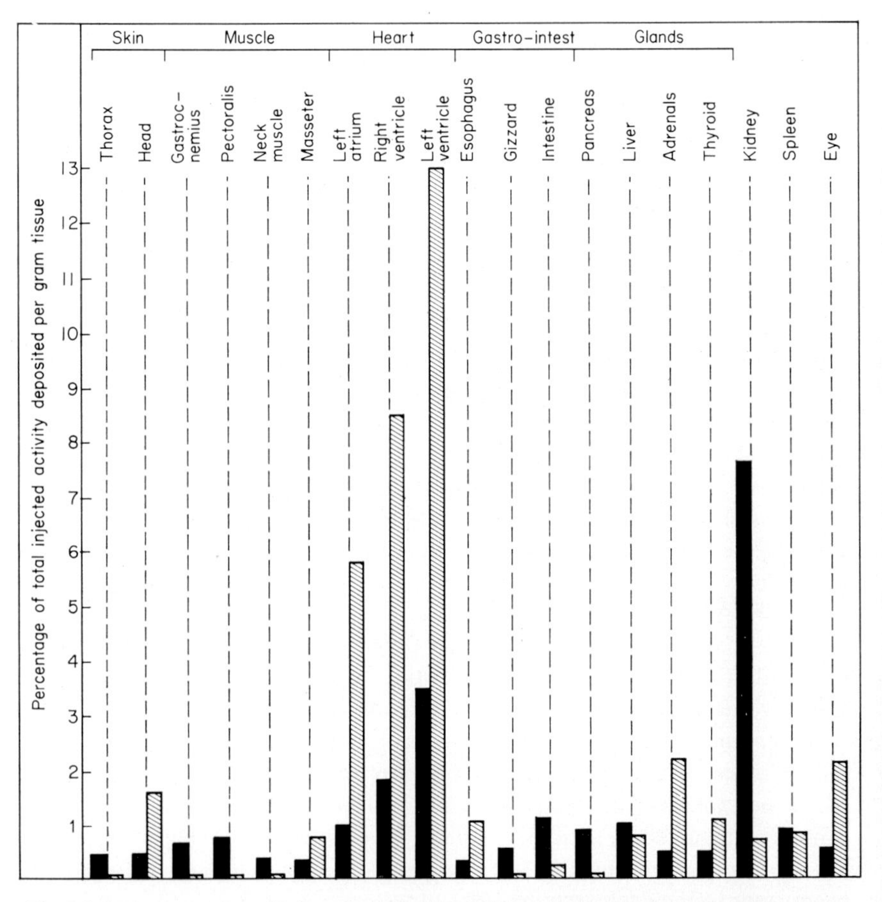

Fig. 7.2. Percentage of total injected activity of Rb^{86} deposited per g tissue in various organs of domestic ducks. The hatched bars indicate values obtained during submersion. The graph represents average values from eight ducks (from Johansen, 1964).

that the lactic-acid concentration of arterial blood of seals increases during the dive. Similar results have been obtained on all divers investigated and pH values of arterial blood below 7·0 are not unusual. Thus even the heart and

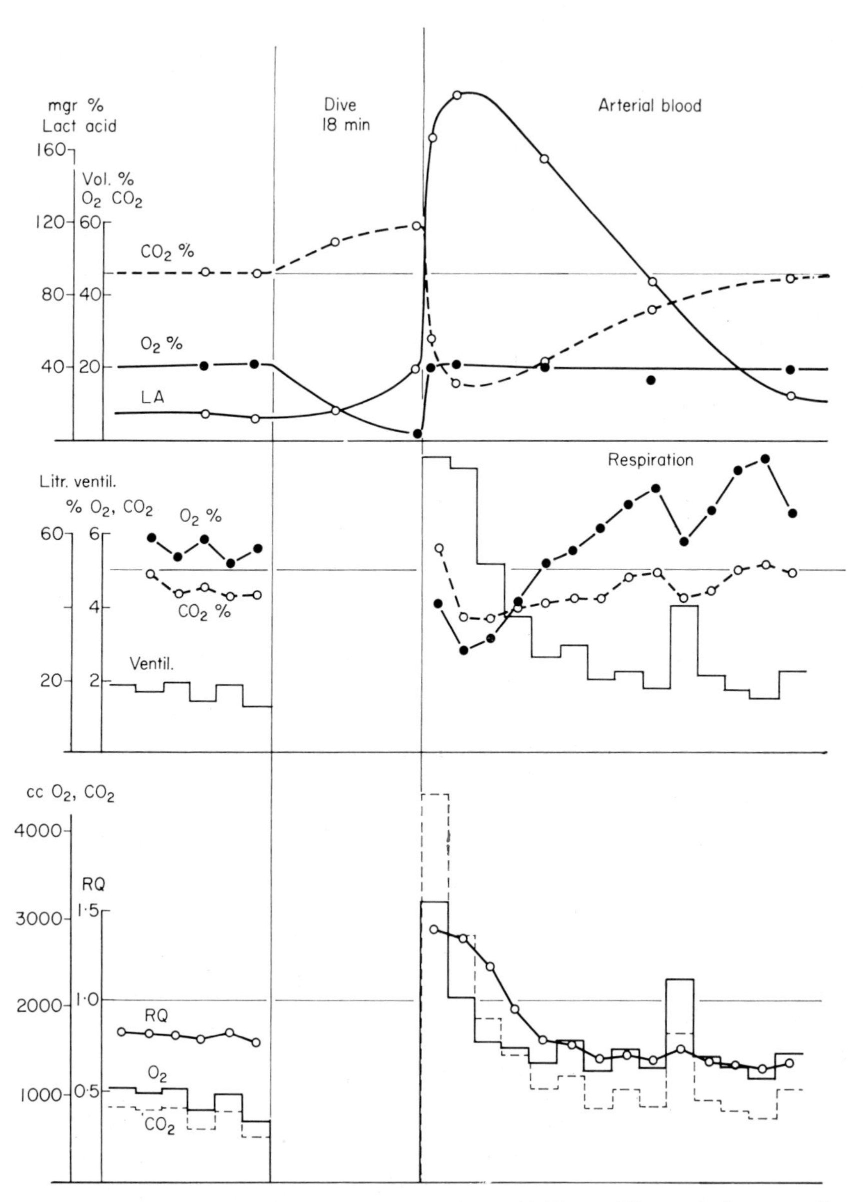

Fig. 7.3. Typical results from diving experiment in a seal. The middle graph shows ventilation in litre/min and the percentages of O_2 and CO_2 in expired gas. The lower graph shows O_2-uptake and CO_2-output in litre/min (from Scholander, 1940).

the brain are exposed to severe acidosis during diving. The non-circulated organs accumulate metabolites during the submersion. Following the dive there is a large increase in the lactic acid content of the blood (Fig. 7.3).

The strict economy with the O_2-stores during the dive is paralleled by a reduced total energy production. This holds true for frogs, toads, alligators, turtles, ducks and seals, and appears therefore to be a general property of the diving response (Andersen, 1966).

Some turtles (for example *Sternothoerus merior*) are able to survive 12 h of submersion. It is most unlikely that aerobic processes play a significant role in energy production during such conditions. Belkin (1962) showed that if glycolysis is inhibited by iodacetate the turtles would survive submersion for only half an hour. This is undoubtedly the most extreme example of anaerobic energy production among divers. It is possibly also an outstanding demonstration of tolerance to anaerobiosis since both brain and heart tissue under these conditions must rely almost entirely upon anaerobic processes.

5. CHEMICAL SENSITIVITY OF VENTILATION

In vertebrates acidosis and hypoxia stimulate ventilation. *A priori*, one would expect divers to be less sensitive to these stimuli, at least during the dive, than non-divers. This problem has received much attention (Andersen, 1966). It appears safe to conclude, at least with regard to seals and ducks, that their ventilatory minute volume is less sensitive, although not insensitive, to changes in arterial P_{CO_2} than that of non-diving vertebrates. Ventilation of divers is also sensitive to hypoxia, but a quantitative comparison with the response of non-divers has not been made.

Chapter 8

PLACENTAL GAS EXCHANGE

The respiratory problems presented by the mammalian foetus are different from those of the bird embryo. In the latter the problem is to utilize the fixed-exchange possibilities of the shell, in the former the foetus respires by means of a living organ, the placenta, that develops in accord with the foetus itself. And, while the outer allantoic capillaries of the bird embryo serve gas exchange only, the placental capillaries have many functions besides gas exchange. As a matter of fact the exchange of gases across the placenta seems to present far fewer problems than the exchange of other substances. Thus Faber and Hart (1966) showed that the exchange of O_2 across the rabbit placenta is entirely flow limited, i.e. the diffusion process is not a limiting process in the overall exchange. Symptomatically the tissue separating the two organisms does not appear to have the structural characteristics of specialized exchange membranes. The subject of placental gas exchange is admirably reviewed by Metcalfe *et al.* (1967).

A. Structure

Gas exchange across the mammalian placenta occurs between maternal and foetal capillaries. The relative geometry of the two systems is therefore of obvious importance to the process. Unfortunately this geometry is not a very orderly one like that of fish gills or of the rete mirabile of the teleostean swimbladder. Likewise the parameters of gas exchange are variable and non-uniform. This applies to the rate and distribution of placental blood flow, to the thickness of the diffusion barrier as well as to the effect of the Bohr-shift. The two sets of placental capillaries are arranged in different ways depending on the species. In the rabbit and sheep Mossman (1926) and Barcroft and Barron (1946) found that the blood streams form a counter-current pattern (Fig. 8.1). This morphological evidence was supported by physiological measurements of blood P_{O_2} in rabbits by Barron and Battaglia (1956), who found a higher P_{O_2} in foetal blood leaving the placenta than in maternal blood leaving it. In the sheep physiological measurements threw doubt on the functional significance of counter-current exchange. Metcalfe *et al.* (1965) found no effect on the rate of gas exchange by reversing the direction of foetal placental blood flow.

In the human placenta the foetal capillaries appear to form loops which pass through blood lacunae perfused by maternal blood. This system has been termed a "pool system" (Fig. 8.2). Some workers prefer to consider it as a "multivillous" system (Fig. 8.3). The functionally important point of this anatomical arrangement is that the maternal blood flows successively past a large number of foetal villi.

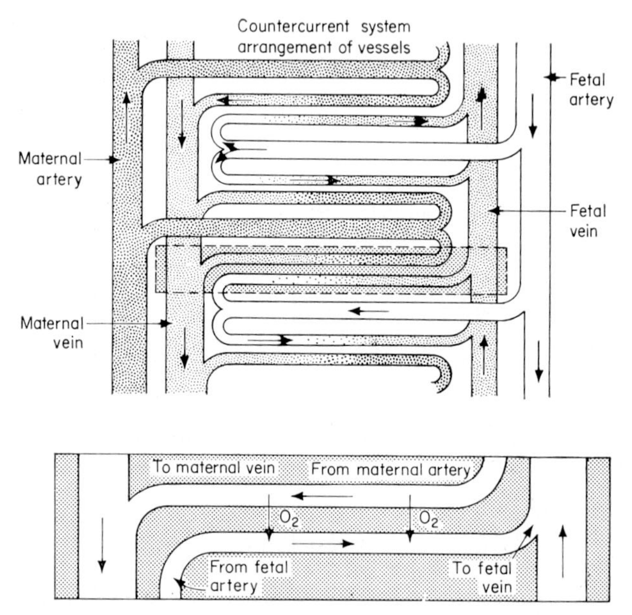

Fig. 8.1. Schematic presentation of pattern of maternal and foetal blood flow during gas exchange in a "counter-current" arrangement of vessels. Functional unit for gas exchange is indicated in upper portion of diagram by stippling and is enlarged in lower portion. Oxygen concentration is represented by intensity of shading (from Metcalfe *et al.*, 1967).

B. Gas Exchange

On theoretical grounds Metcalfe *et al.* (1967) reasoned that of these systems the counter-current arrangement would be about twice as efficient as a con-current system, while the multivillous system is of intermediate efficiency. However, diffusion capacities, i.e. the O_2-flux per mm Hg ΔP_{O_2} between maternal and foetal blood in the placenta from animals with these different capillary arrangements, did not indicate higher values for one type than for any other. Neither did the extraction of O_2 from maternal blood vary according to the geometry.

The thickness of tissue separating maternal from foetal blood is very variable even within one placenta. Bartels and Metcalfe (1965), reviewing the subject, suggested that some parts of the placenta are concerned primarily with gas exchange while others are concerned with other processes requiring more tissue substance.

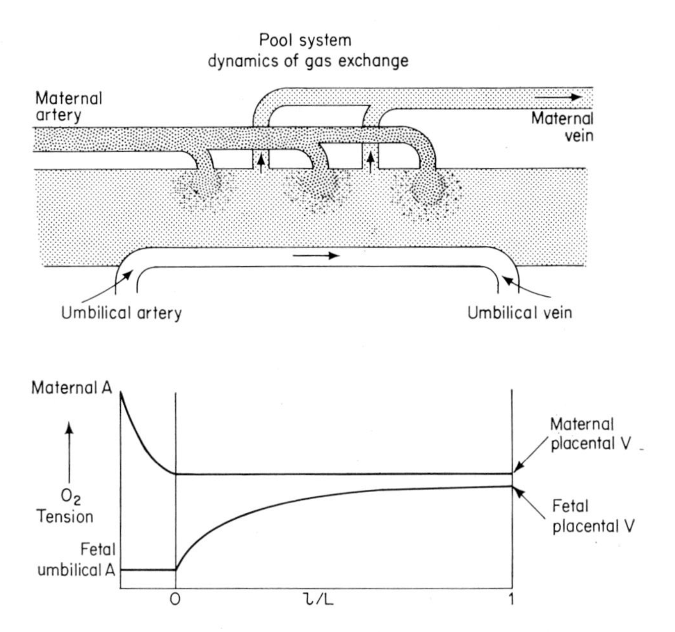

Fig. 8.2. Schematic presentation of the dynamics of gas exchange in a "pool" system. Foetal blood is exposed during its passage through the villar capillary to maternal blood of a uniform oxygen tension represented by that found in maternal placental venous channels. The abscissa represents the length of the exchange channel (from Metcalfe *et al.*, 1967).

C. Blood Flow

The blood flow to different parts of the placenta varies with time and a considerable, although varying proportion of both maternal and placental blood is shunted past the exchange capillaries. The uneven blood distribution was clearly demonstrated in the sheep placenta by Power *et al.* (1967) who measured the distribution of radioisotopes in the placenta after injection of the tracer in the maternal and placental blood supply. Longo *et al.* (1968) also demonstrated large variations in the P_{O_2} in blood samples from maternal and placental venules (Fig. 8.4) in the dog placenta.

D. Blood

Among water breathers it appears to be a general rule that animals living in O_2-poor environments have blood with a low P_{50}, i.e. high O_2-affinity. The same relationship exists between the blood of an adult and a foetus of the same species. For most of its 22–23 month gestation period the spiny dogfish *Squalus suckleyi* has a foetal Hb with higher O_2-affinity than the adult pigment (Manwell, 1958). Similarly, metamorphosis of the post-larva of the teleost *Scorpacnichtys* is accompanied by a change in Hb type just as in the metamorphosis of the bullfrog. An oviparous ray has a transient embryonic Hb of high O_2-affinity during the first months of development (Prosser and

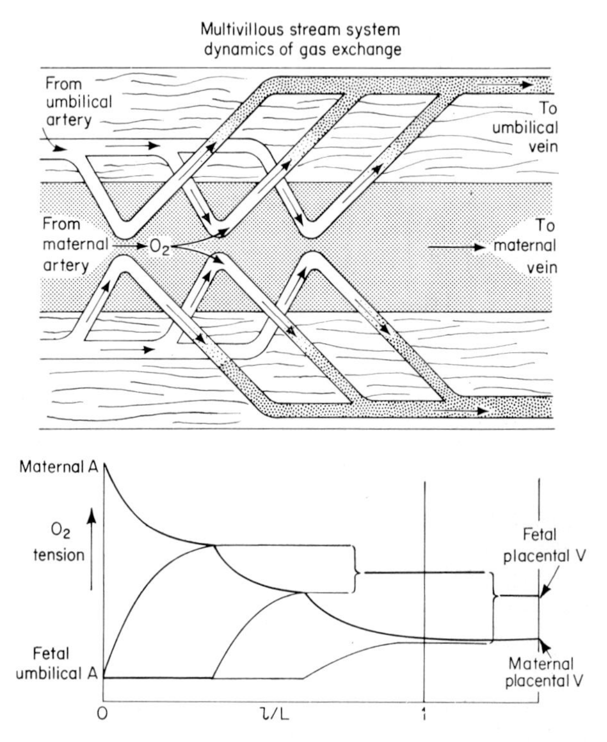

Fig. 8.3. Gas exchange in a "multivillous stream" vascular pattern. This scheme shows that blood in different foetal capillaries is exposed to maternal blood with different oxygen tensions. The blood within each foetal capillary approaches the oxygen tension of the surrounding maternal blood. Foetal blood leaving the placenta represents a mixture of blood from different capillaries and its oxygen tension may exceed that in maternal blood leaving the placenta. *Abscissa* represents the distance travelled (I) by all the foetal blood in the diagram from the beginning of gas exchange in upstream capillary along entire length (L) of the gas exchange unit to the exit of downstream capillary from maternal intervillous space (from Metcalfe *et al.*, 1967).

Brown, 1962). In all mammals investigated except the cat (Novy and Parer, 1969) the foetal blood has higher O_2-affinity than the maternal blood, also under *in vivo* conditions (Table 8.1). The blood of the bird embryo also has higher O_2-affinity than that of the adult (Bartels *et al.*, 1966).

The O_2-capacity of embryonic and adult blood does not show a similarly regular pattern. The O_2-capacity of the unhatched chicken is below that of the hen (White, 1968). In mammals the foetal O_2-capacity may be below, equal to or above the maternal. Table 8.1 shows respiratory characteristics of foetal and maternal blood from several species.

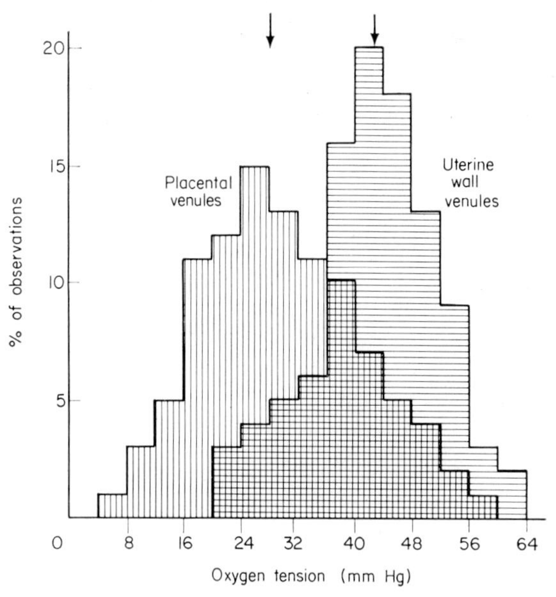

Fig. 8.4. Histogram showing the range of P_{O2} in placental venules and uterine wall venules. This demonstrates the different degree of equilibration in the different parts of a placenta (from Longo *et al.*, 1968).

TABLE 8.1

Oxygen-affinity (P_{50}) and oxygen-carrying capacity of maternal and foetal blood of various species (from Novy and Parer 1969).

Species	P_{50} at pH 7·40 mm Hg		O_2-capacity ml O_2/100 ml	
	Maternal	Foetal	Maternal	Foetal
Man	26	22	15	22
Rhesus monkey	32	19	15	18
Rabbit	31	27	15	14
Sheep	34	17	15	17
Goat	30	19	13	12
Pig	33	22	13	13
Elephant	24	21	20	17
Camel	20	17	15	17
Llama	21	18	14	19
Cat	36	36	12	16

Bartels and Metcalfe (1965) discuss the effect of the different respiratory properties of foetal and maternal blood upon placental gas exchange. The subject is complicated by the fact that the effect of these properties depends on several other parameters, like relative blood flow and diffusion characteristics of the placental barrier.

The foetal blood transports O_2 from the placenta to the tissues of the foetus. Its respiratory properties must therefore represent an optimal compromise between O_2-delivery to the tissues and O_2-uptake in the placenta. The obser-

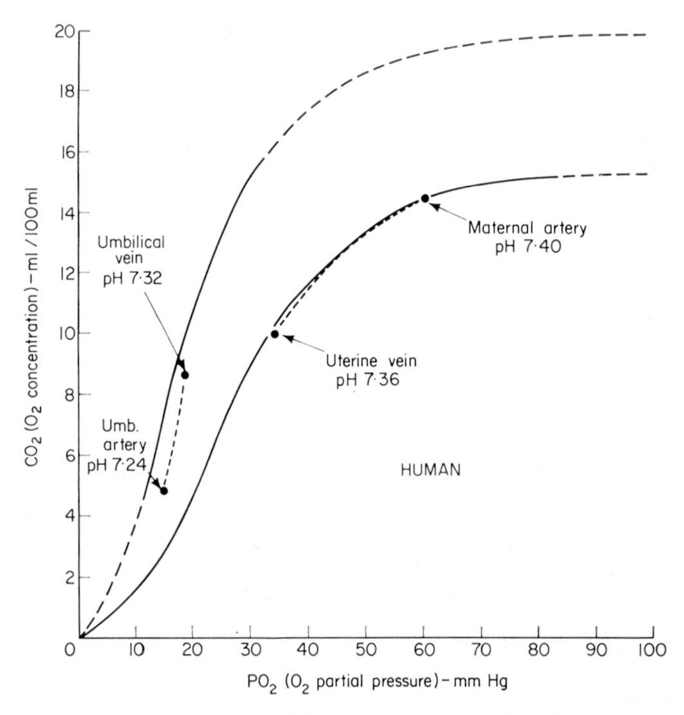

Fig. 8.5. Representative values reported for oxygen concentrations in blood samples from uterine and umbilical vessels are placed on oxyhaemoglobin dissociation curves appropriate to the pH of each blood sample. During placental gas exchange, oxygen tension in maternal blood falls along the dotted line on the right, departing from the dissociation curve for pH 7·4 as the blood pH falls, due to the influx of carbon dioxide and fixed acids from foetal circulation. Reciprocal changes occur in foetal blood during its passage from the umbilical artery to umbilical vein, the oxygen tension and concentration rising along dotted line on left (from Metcalfe *et al.*, 1967).

vation that the P_{O_2} in the foetal blood leaving the placenta is rarely much higher than the maternal venous placenta blood, signals that the placenta is not particularly specialized for gas exchange. The low P_{50} still allows the blood to be well saturated at this low P_{O_2}. In this way gas exchange in the foetal tissues will occur in a P_{O_2} range corresponding to the steep portion of the O_2–Hb curve (Fig. 8.5). Thus the low foetal P_{50} is an adaptation of the poor diffusion properties of the placenta in much the same way as the low P_{50} of some crustacean bloods appears to be an adaptation to a poorly permeable gill barrier. It may be significant that in the rabbit there is a rather efficient counter-current gas exchange in the placenta, and a small difference between P_{50} of foetal and maternal blood.

A high foetal O_2-capacity compared to the maternal helps to put foetal gas exchange in the steep P_{O_2} range of their O_2-dissociation curve. It may also have significance in reducing the foetal cardiac output, thus saving energy.

A favourable transplacental gradient is further secured by the Bohr-effect which is present in both types of blood. Thus in the placenta CO_2-flux will cause acidification of the maternal blood which increases its P_{O_2} whereas the foetal blood becomes more basic whereby the P_{O_2} decreases. This is also shown in Fig. 8.5 where the *in vivo* points of maternal and foetal blood leaving the placenta are farther apart than they would have been had no Bohr-effect been present. The beneficial effect of this double Bohr-effect is further enhanced by the fact that CO_2 diffuses faster than O_2. Although the rate of the Bohr-shift of foetal blood has not been measured it appears reasonable that it should have about the same half time as the maternal Bohr-effect which is about 0·12 sec (Forster and Steen, 1968). Thus the rapid CO_2-flux will tend to complete the decrease of the maternal pH and the increase of the foetal during the early period of exchange whereby O_2-exchange will occur at larger ΔP_{O_2} values.

Chapter 9

GAS EXCHANGE OF THE BIRD EGG

The fact that the bird foetus is enclosed in a rigid shell of constant area and thickness throughout a developmental period that entails 1000-fold increase of gas exchange (Fig. 9.1) gives it a quite unique position among respiring

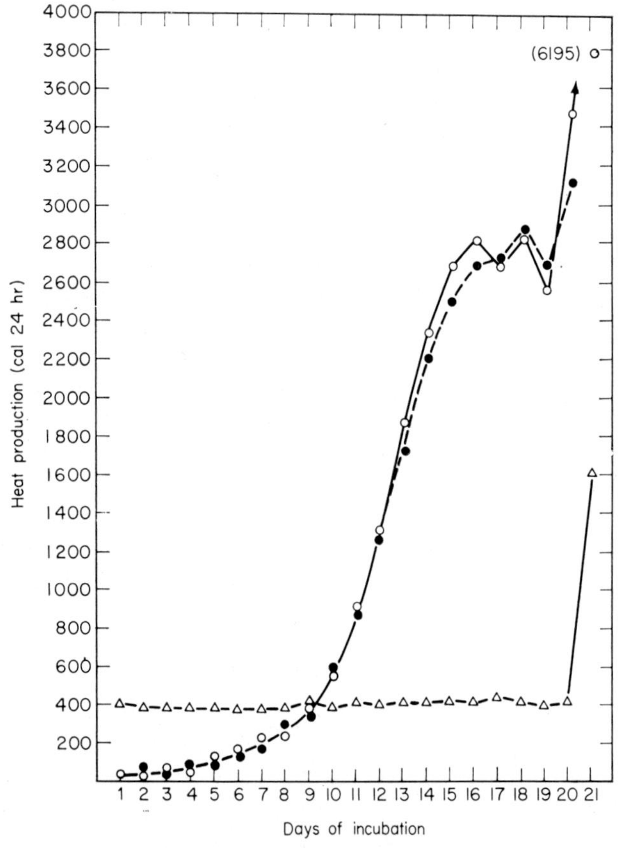

Fig. 9.1. Heat production of the growing chicken embryo in one egg (64 g, Blue North Holland), estimated by direct (●) and indirect (○) calorimetry; △, evaporative heat loss. Temp. = 37·7°C; relative humidity = 10–28 per cent (from Romijn and Lokhorst, 1960).

organisms. The following account will discuss how this increase in capacity is achieved and how it is compatible with restricted water loss. We shall start with a description of the anatomy of the egg shell and then briefly review the development of the chick before discussing the rather limited information on the respiratory process itself.

The anatomy of the egg is thoroughly reviewed by Romanoff and Romanoff (1949). Most bird eggs are covered by an external cuticle. This is a non-cellular protein (probably mucin) coat, which is securely attached to the shell, and covers the entire surface of the egg without detectable pores. The thickness

TABLE 9.1

The weight and shell thickness of eggs from several species of birds (from Romanoff and Romanoff, 1949).

Species	Egg weight (g)	Shell thickness (mm)
Aepyornis	12000	4·40
African ostrich	1400	1·95
Australian swan	700	0·69
Holland turkey	80	0·41
Chickens:		
Cochin China	65	0·36
Leghorn	58	0·31
Bantam	38	0·26
Ringnecked pheasant	32	0·26
Quail	9	0·13
Australian finch	1	0·09
Hummingbird	0·5	0·06

of the cuticle varies markedly from species to species and in some cases also from area to area on the same egg. In the chicken egg, which we will be mostly concerned with, it is from 5 to 10 μ thick. Beneath the cuticle is the egg shell proper which is a hard, calcareous and porous structure. The thickness of the egg shell varies, first of all among species but also according to the nutritional and external conditions of each individual. As shown in Table 9.1 the shell becomes progressively thinner the smaller the egg.

We are aware of many functions that the egg shell fulfills. It must be a rigid box protecting the embryo both from mechanical damage and from infections. It must supply the embryo with sufficient amounts of inorganic salts during its development. It must allow diffusion of respiratory gases while still restricting water loss. The shape and structure of the shell ensures its mechanical functions, its composition and micro-architecture the exchange functions.

The egg shell is not a homogeneous structure. It is composed of an organic matrix or framework, which makes up about 2% of the total shell weight, and of an interstitial substance composed of a mixture of carbonates and phosphates of calcium and magnesium. The matrix is a collagen-like protein. The outer part of the egg shell is a rather spongy layer, penetrated by numerous pores. Their opening to the exterior is covered by the cuticle, while their inner portion is continuous with the numerous air channels in the inner or mammillary layer of the egg shell. From our point of view the pores have special interest, they are usually most abundant at the blunt end of the egg. The frequency of pores varies among species; in normal chicken eggs there may be a total of about 7500 pores, with a frequency of about 90–130 per cm^2. Usually the pores have an oval cross-section. Table 9.2 gives an idea of the size of the pores from the eggs of various birds. Hens' eggs have oval pores with the longest diameter about 25 μ and the shortest about 10 μ. The pores open to the exterior, not only through the area of the pore itself but it widens out considerably ending in a set of grooves in the outer surface of the egg shell. This increases the diffusion area between the external atmosphere and the air in the mouth of the pore itself.

TABLE 9.2

The dimensions and shape of egg-shell pores in eggs of several birds (from Romanoff and Romanoff, 1949).

Species	Largest pore (mm)	Smallest pore (mm)	Shape
Ostrich	0·050	0·020	Circular
Swan	0·042 × 0·038	0·029 × 0·026	Oval
Turkey	0·055 × 0·037	0·037 × 0·031	Oval
Duck	0·036 × 0·031	0·014 × 0·012	Oval
Hen	0·029 × 0·022	0·011 × 0·009	Oval
Pheasant	0·014 × 0·012	0·013 × 0·010	Oval
Auk	0·090 × 0·057	0·011 × 0·006	Oval
Gull	0·016 × 0·013	0·011 × 0·010	Oval

The contents of the egg are separated from the shell by two membranes, the innermost egg membrane and the outermost shell membrane. The inner membrane surrounds the albumen and the outer is firmly cemented to the shell. The two membranes are attached to each other along the entire area of the egg, except in a small area at the blunt end where the air space is located. Table 9.3 gives the thickness of the two membranes in a few species of birds. This parameter varies not only among species but also within the same species. The thickness may also vary from one area to another of the same egg.

Both membranes are composed of branched or unbranched keratin and mucin fibres that form a net-like structure. The interstices between them are filled with an albuminous cementing material, which is particularily abundant at the smooth face towards the albumen.

It is stated by Hays and Sumbardo (1927) that both these membranes have pores which are particularly numerous in the inner membrane. The pore openings may be as large as 30 μ. Whether these pores are filled by fluid or gas, is a question which we will discuss later.

As already mentioned the two membranes are separated by an air space at the blunt end of the egg. In a freshly laid egg there is no such air space; it develops during the first hours after laying. In the beginning it is rather small, but grows with age. The volume of the air space during development has been investigated by Romijn and Roos (1938). In White Leghorn hens they found the air space in the beginning to be 0·4 cc while at the end of incubation it was 12 cc in volume (Fig. 9.2). The air space serves two functions, one is to take up the space produced as a result of water loss, the other to provide a respiratory environment for the embryo at the later stages of its development.

TABLE 9.3

Thickness of the egg membranes of egg of some birds (from Romanoff and Romanoff, 1949).

Species	Thickness of membranes (mm)
Ostrich	0·200
Swan	0·165
Chickens:	
Brahma	0·092
Leghorn	0·065
Bantam	0·050
Quail	0·067
Zebra finch	0·005

The embryo develops inside this four-layered structure. In the beginning it has no respiratory system, and barely a circulatory one. Most likely it obtains its oxygen from the surrounding egg yolk and albumen which exchanges gas through the surface of the egg. As development proceeds the allantoic sac develops. Its lumen acts as a reservoir for excretory products, while its outer surface is well vascularized and, as development proceeds, becomes situated next to the egg membrane. This outer area of the allantoic sac is the respiratory organ of the developing embryo. At the end of development the allantois covers the entire surface of the inner membrane. The thickness of the wall between blood in the allantoic capillaries and the egg membrane is exceedingly small, less than 1μ. The degree of capillarization of the outer allantoic surface is increased if incubation occurs at subnormal P_{O_2} (Remotti, 1933).

The development of the chick from a few cells to a full-grown chicken ready to hatch implies a greatly increased demand for gas exchange (Fig. 9.1). This increase is paralleled by changes in the structure of the egg. For one thing the allantoic membrane increases in area so that an increasing proportion of the egg shell comes into play as development proceeds. Also the air space increases in volume and gas exchange between the underlying allantoic membrane and this air space is probably more efficient than that through the main part of the egg shell.

Gas exchange during the later stages of development in the chicken has been investigated by Visschedijk (1968a, b and c). Towards the end of incubation the lungs of the young bird itself have developed sufficiently to assist in

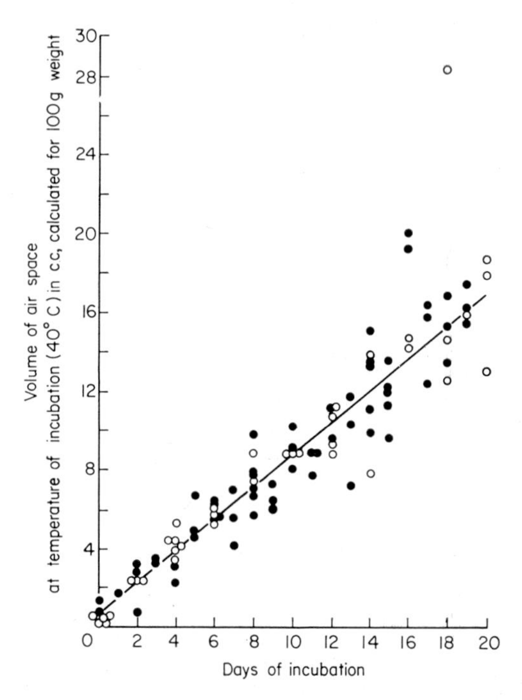

Fig. 9.2. Increase of the air space during incubation of fertile eggs (●), and unfertilized eggs (○) during incubation (from Romijn and Roos, 1938).

respiration. To accomplish this the chicken makes a hole through the allantois beneath the air space and ventilates with this gas. In the beginning ventilation is weak and irregular, and contributes a rather small portion of the total gas exchange. However, as hatching approaches an increasing proportion of the total gas exchange occurs between the air space and the lungs of the chicken. This is not to say that gas exchange between the allantoic membrane surrounding the rest of the chicken ceases during this period. This occurs first after the chicken has pipped the shell, that is, made the first hole, and from this moment on the allantoic sac dries up and rapidly loses its circulation and thereby its respiratory function. At the same time the air space comes into direct contact with the atmospheric air which enhances gas renewal in it and lung respiration by the chick rapidly becomes the dominant respiratory process. This continues for another day or so, at which time the chicken starts to break open the shell. This evidently is initiated by the high P_{CO_2} and low P_{O_2} in the air space. The act of hatching requires vastly increased gas exchange which is possible only by direct ventilation with the atmospheric air. These events are illustrated in Fig. 9.3.

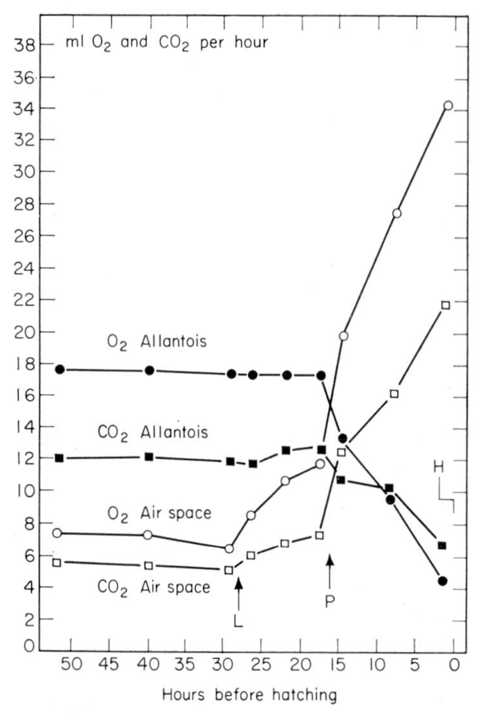

Fig. 9.3. The mean gaseous exchange through the shell over the space and allantoic shell, plotted against time. L = onset of lung ventilation; P = pipping; H = hatching (from Visschedijk, 1968).

The egg is one of the few biological structures that has a simple enough architecture so that we can describe its diffusion properties somewhat quantitatively:

We can measure the area and thickness of the diffusion barrier with some precision, as well as the P_{O_2} in ambient air and in the air space. White (1968) measured O_2-saturation in blood to and from the allantois in a 15-day-old chick embryo. The arterial blood showed 32% saturation, the venous allantoic 65%. The blood from the allantois is mixed with venous blood from the yolk sac and liver before it enters the heart. The systemic arterial blood is therefore about 50% saturated. It is rather surprising that such a degree of hypoxia is a normal state for the embryo. The O_2-dissociation curve of embryonic blood at this stage also shows adaptation to such conditions. Bartels *et al.* (1966) found P_{50} of about 30 mm Hg in 20-day-old embryos as compared to a P_{50} of 50 mm Hg a few days after hatching. In this respect the chick is similar to most mammalian embryos.

Kutchai and Steen (in press) investigated the permeability of the egg shell and egg membranes during this development. The gas permeability of a newly laid egg is too low to permit the gas exchange which occurs towards the end of development. Measurements showed a marked increase in gas permeability during the first 5–6 days of incubation, after this time it increased only a little (Fig. 9.4). This increase was due to a gradual drying out of the membranes (Fig. 9.5). Neither the increase in gas permeability nor the reduced water content was found in unfertilized eggs incubated together with the fertilized ones.

There is evidence to suggest that the reduced water content of the egg membranes is due to an increased colloid osmotic effect of the albumen of fertilized eggs.

The CO_2 permeability of the egg shell plus egg membranes was at all stages of incubation 2–3 times higher than the O_2 permeability. This suggests that diffusion occurs primarily through air-filled, rather than through fluid-filled, pores. This ratio of permeabilities appears important since it contributes to maintaining in the blood of the chick embryo P_{O_2} and P_{CO_2} values close to those in the chick after hatching. If the diffusion barrier had been water, the permeability to CO_2 would have been about 30 times that to O_2. This would mean that the P_{CO_2} would be one-thirtieth of the P_{O_2} at R.Q. = 1. With an arterial P_{O_2} of 55 mm Hg this would result in an arterial P_{CO_2} of 3–4 mm Hg. This would no doubt cause unpleasant acid-base changes at hatching.

Kutchai and Steen (in press) also measured the fractional pore area and the equivalent pore ratio of both egg membranes. Some anatomical characteristics are presented in Fig. 9.6 together with permeability values.

The egg has the same water conservation problem as a diapausing insect pupa. In the bird egg this problem is further complicated by the fact that the gas exchange increases by a factor of 1000 during the time of incubation. In most animals water loss parallels gas exchange. In bird eggs this is not the case (Fig. 9.2). The situation evidently develops in the following way: in a newly laid egg the shell is saturated with water and covered by the cuticle. Water loss is thus limited not by evaporation from the egg surface, which, because of the cuticle is dry, but by diffusion between water inside the cuticle and air outside it. As development proceeds, the egg shell dries up and the pores are progressively filled by more and more air, causing the

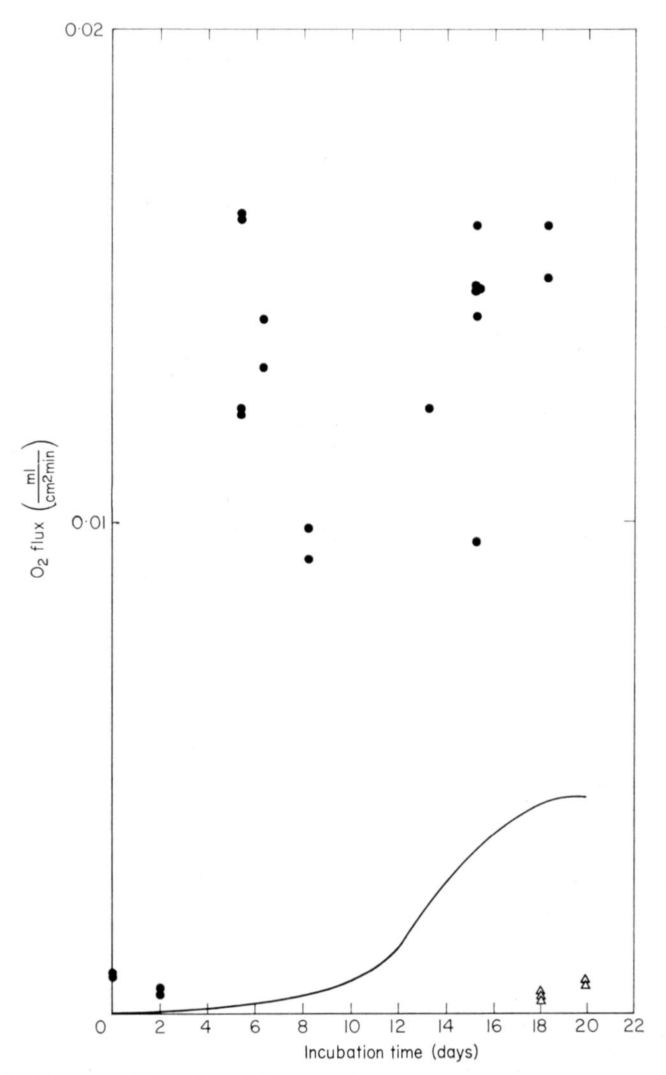

Fig. 9.4. O_2 flux caused by a P_{O_2} difference of 155 mm Hg across shell plus membranes of fertile eggs (●) and infertile eggs (△) as a function of incubation time. The solid curve shows the O_2 uptake rates observed in fertile eggs by Romanoff (from Kutchai and Steen, in press).

increased gas permeability. However, water loss is still limited by diffusion across the cuticle.

In certain respects this resembles the way water loss is limited in insect pupae. The general principle applied in both instances is to limit water loss, not by reducing the evaporative surface, which in most cases is equal to the respiratory surface, but to limit it by diffusion through a limited area.

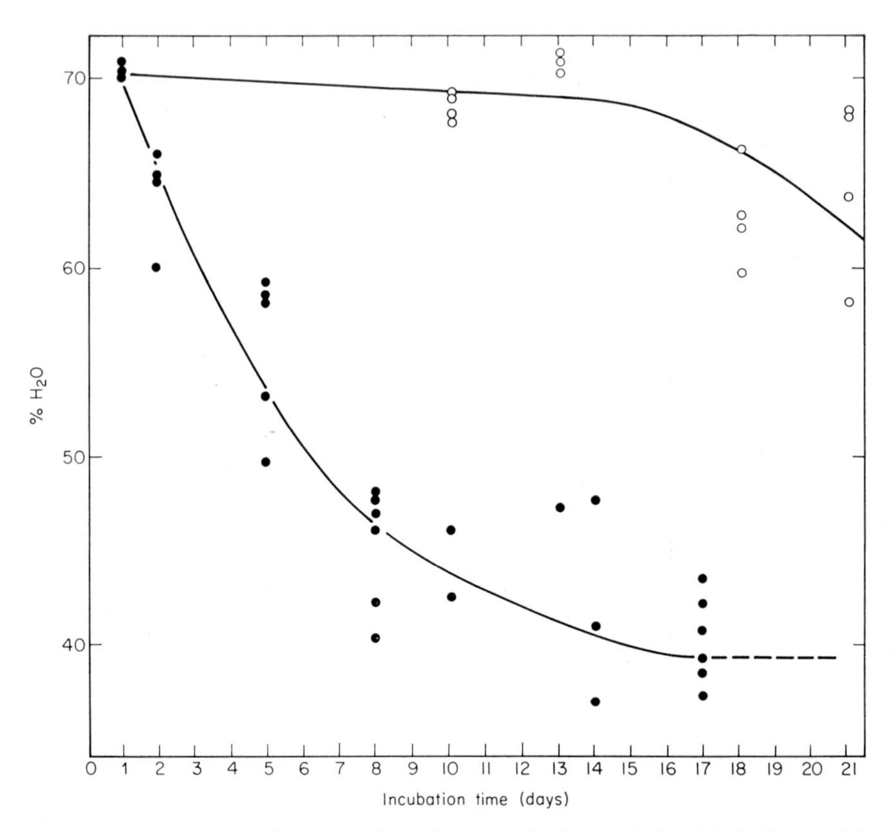

Fig. 9.5. Water content of compound membranes of fertile eggs (●) and infertile eggs (○) as a function of incubation time (from Kutchai and Steen, in press).

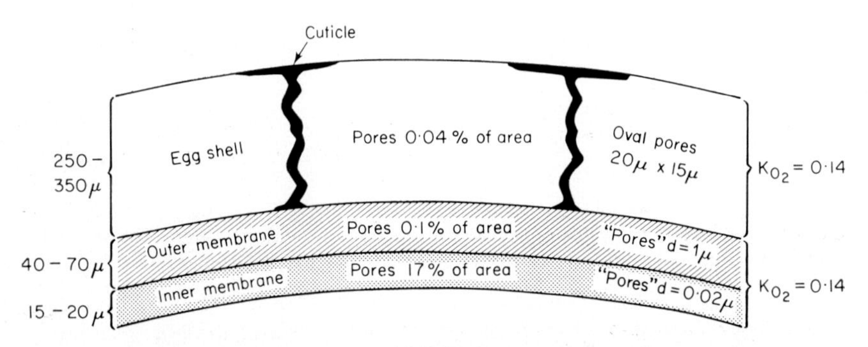

Fig. 9.6. Diagrammatic representation of the shell and shell membranes of a hen's egg showing structural features and O_2 permeabilities. O_2 permeability (K_{O_2}) values are in ml (STP)/(cm^2 min atm) and are intended to apply to fertilized eggs that have been incubated 5 days or longer (from Kutchai and Steen, in press).

Chapter 10

THE SWIMBLADDER FUNCTION

The swimbladder is of interest not so much because it may serve as an auxiliary respiratory organ during certain conditions of limited external O_2-supply, but rather because it exhibits a unique interaction between respiratory parameters.

The outstanding property of the swimbladder is to produce a P_{O_2} inside the bladder that is hundreds of times higher than that present in the arterial blood which enters it. This is achieved, not by active transport of anything across cells, but simply by combining the properties of haemoglobin as an O_2-carrier with the diffusion laws in a particular circulatory arrangement.

The swimbladder is a gas-filled organ. It receives blood from an artery which forms a capillary net at a localized area of the bladder or on the inner surface of it. The capillaries are drained through a vein which runs parallel to the artery. A certain distance from this capillary system both artery and vein divide into another system of capillaries which run parallel for 1–20 mm (depending on the species) before they reunite (Fig. 10.1). The capillary system of arterial and venous origin intertwine in such a way that arterial capillaries are surrounded primarily by venous ones and vice versa (Fig. 10.2). This organ is called a rete. The anatomist who first recognized its unique architecture called it rete *mirabile*. Mirabile, which means wonderful, has been dropped, possibly because the organ looks too modest to deserve such an adjective. Personally, I prefer to maintain the term.

The architecture of the rete has two important properties. One is that the arterial and the venous blood are brought into an optimal diffusion situation relative to each other; the other is that they flow in opposite directions. The exchange of material between the two blood streams is therefore called counter-current exchange. In the rete of the common eel Krogh (1922) counted about 100 000 capillaries of each type. The distance between arterial and venous blood is about 1 μ and the entire organ has a diameter of 3–6 mm. The aggregate length of capillaries was 900 m, while their area was 210 m^2. No other organ exhibits a larger ratio of exchange area to volume. The rete is therefore the most specialized exchange organ known.

The counter-current flow has the important consequence that diffusible material theoretically can exchange almost completely (Fig. 10.3). In the swimbladder of deep-sea fishes the P_{O_2} may be 100 atm while the P_{O_2} of arterial blood rarely exceeds 0·20 atm. Thus under such circumstances the circulation of the bladder would remove gas from the bladder were it not for counter-current exchange in the rete. Scholander (1954) calculated that the rete of the eel was efficient enough to check leakage, even if the bladder P_{O_2} was 3000 atm, provided a reasonable amount of gas was continuously deposited.

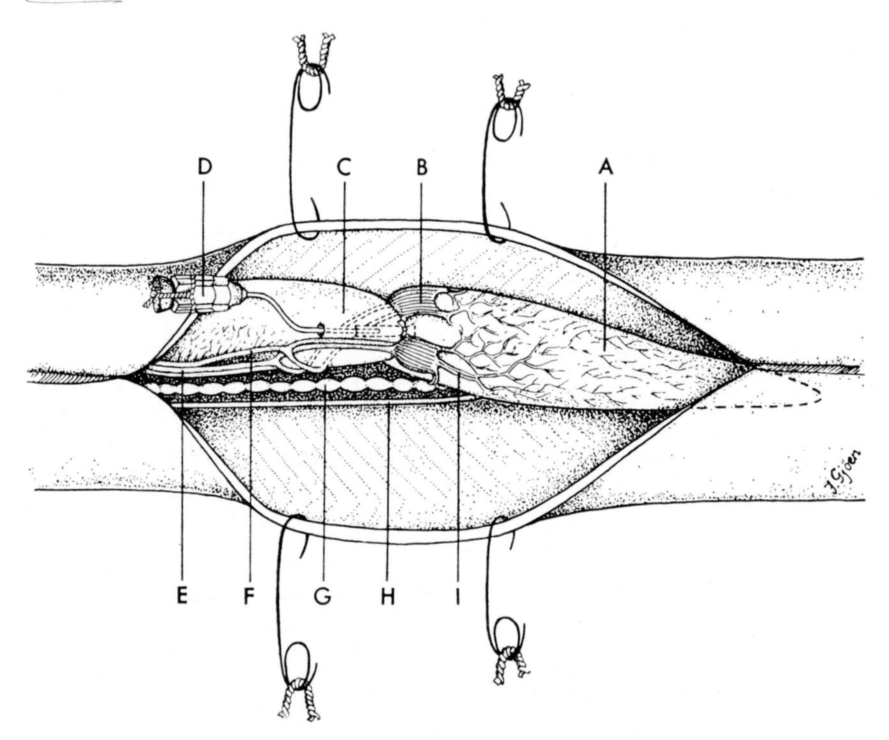

Fig. 10.1A. Shows a drawing (slightly reduced) of the swimbladder of an eel. A modified cannula is inserted into the secretory bladder. Intestines and gonads removed. A = secretory/bladder. B = rete. C = reabsorbent bladder. D = modified cannula. E = pre-rete artery. F = post-rete vein. G = dorsal artery. H = dorsal vein. I = post-rete artery and pre-rete vein (from Steen, 1963b).

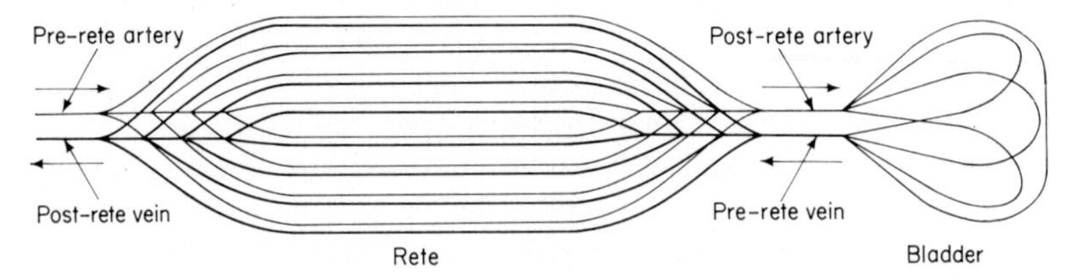

Fig. 10.1B. A schematic drawing of the swimbladder circulation.

What we see in the rete is thus a further development of the counter-current exchange that is present in the gills. There, venous blood could achieve almost the same P_{O_2} as that of the incoming water; consequently a high degree of extraction was achieved. In the rete, arterial blood achieves almost the P_{O_2} of venous blood from the bladder and thus extracts almost all the O_2 which is carried from the bladder.

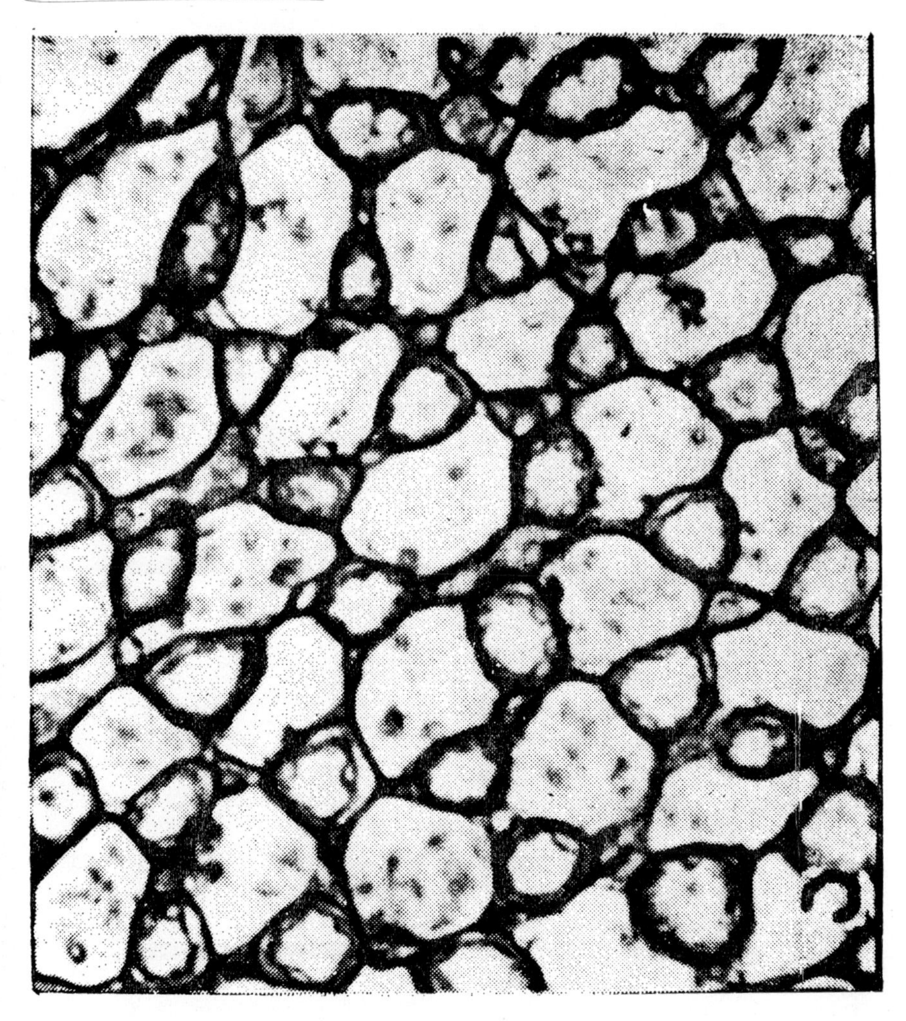

Fig. 10.2. A cross-section of the rete from *Coryphocnoides*. Note triangular arterial and hexagonal venous capillaries. Diffusion path about 1 μ thick (from Scholander, 1954).

However, the ability to preserve high pressures would be of little use, and never recognized, if the system could not build up high gas pressures to begin with. The main problem is therefore: what mechanism operates to fill the bladder with O_2 at a P_{O_2} of 100 atm when it is supplied by blood with no more than 0·20 atm of O_2? The key to the solution of the problem turned out to be the "hairpin counter-current multiplication principle" which had been recognized by engineers long before it was introduced to biology in 1951 by

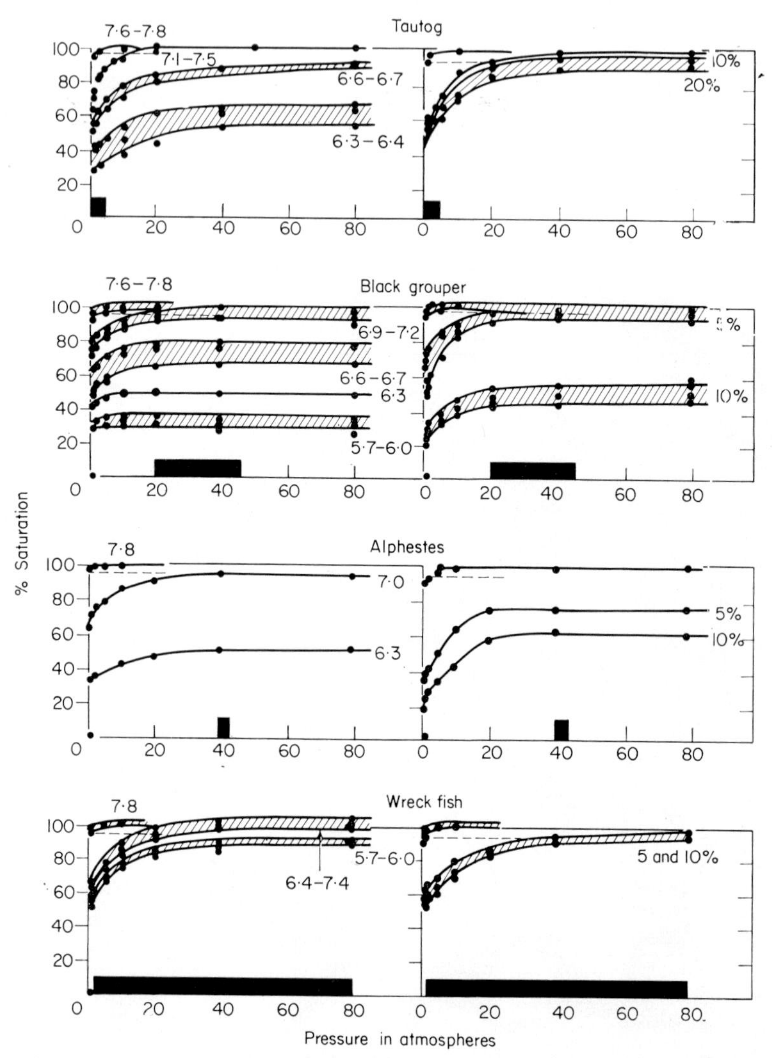

Fig. 10.3. Oxygen equilibrium curves in various fishes at 10°C. Lactic acid was added to the blood in the curves on the left, CO_2 in those on the right. The pH values are given for curves of the lactic-acid series. The equilibration pressure of CO_2 is given as a percentage of 1 atm (from Scholander and van Dam, 1954).

Hargitay and Kuhn in a paper on counter-current multiplication as the mechanism of urine concentration in the loop of Henle. In 1961 Kuhn and Kuhn applied the same principle to O_2-concentration by the rete. And,

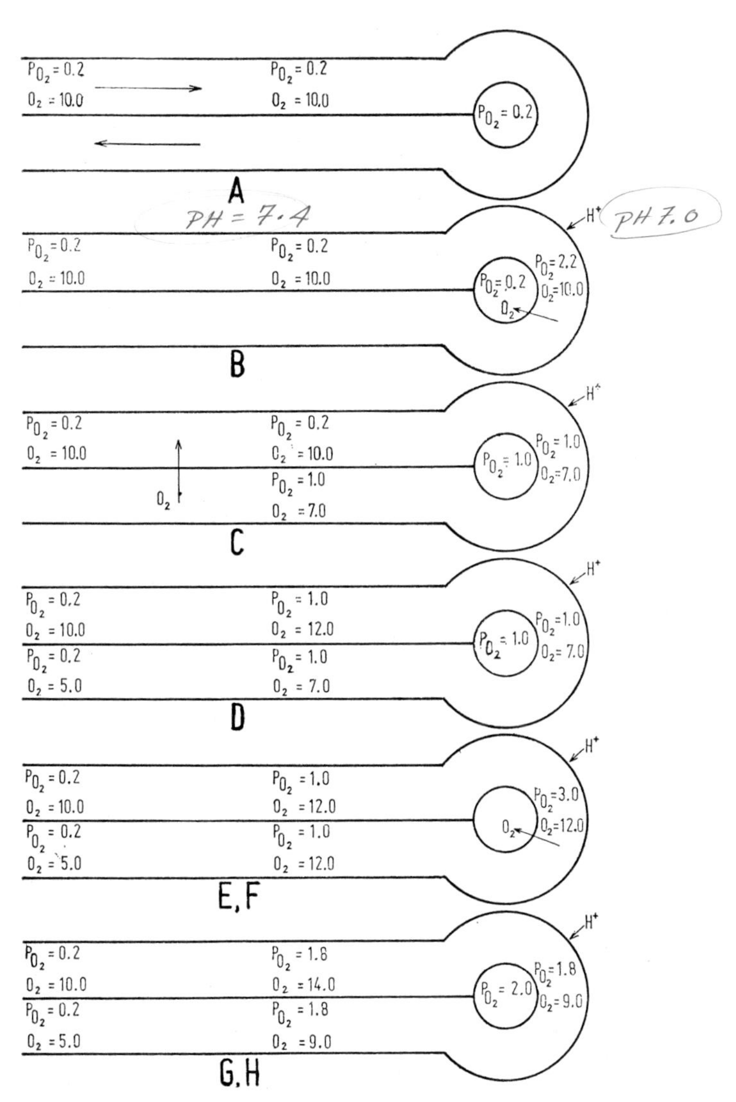

Fig. 10.4. A numerical example of hairpin counter-current multiplication of O_2 by the rete of a fish swimbladder. The same events are also illustrated by Fig. 10.5. The Root-shift is exaggerated. For pedagogical purposes the concentration mechanism is shown as occurring in steps. Time intervals are arbitrary.

although later research has necessitated certain refinements (Berg and Steen, 1968), their model still forms the backbone of the present concept of how the bladder builds up high gas pressures.

The trick which enables the rete to concentrate gases is that the solubility of gases in the blood is reduced as it circulates the bladder. This increases the ratio between gas tension and gas content so that in the rete gases may diffuse from blood with low to blood with high gas content.

A particular reaction of fish blood, the Root-effect, a reduction in O_2-capacity caused by decreased pH (Fig. 10.3), plays an important role in this connection. Scholander and van Dam (1954) found that the reduction in O_2-capacity with reduction in pH persisted in some fishes up to P_{O_2} of 140 atm, while in others it was nullified at lower pressures (Fig. 10.3).

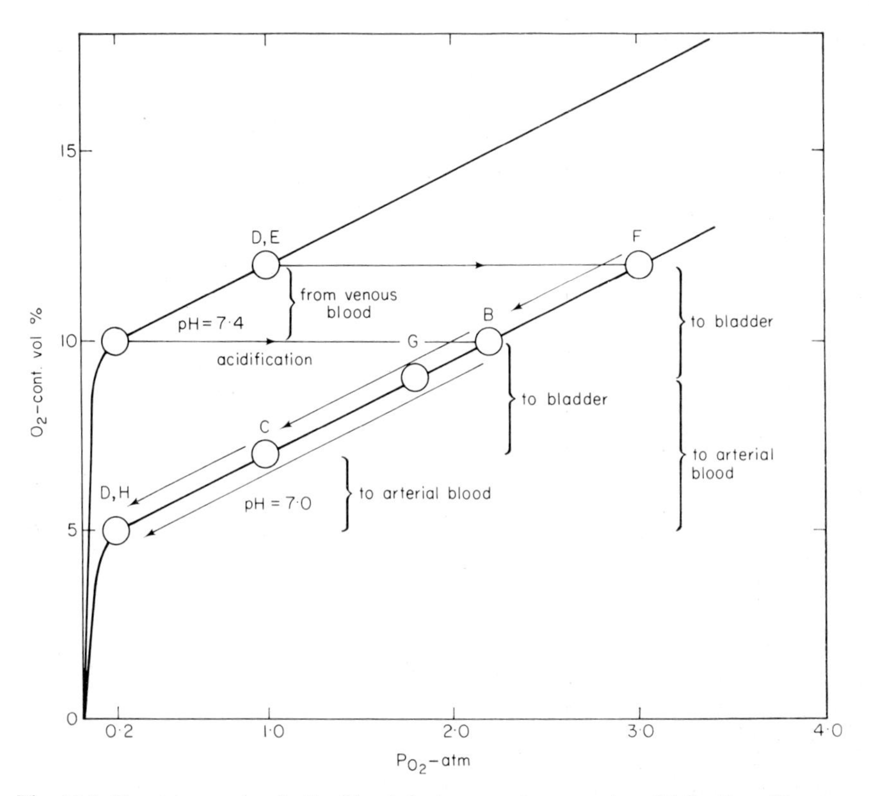

Fig. 10.5. Events occurring in the blood during counter-current multiplication. Compare with Fig. 10.4 A-H.

The following is a simplified and schematic description of how we believe gases are concentrated in the swimbladder. We assume that the rete and the bladder contain no blood to begin with and that the rete is so efficient that no arteriovenous ΔP_{O_2} will exist. As circulation starts (Figs 10.4 and 10.5) arterial blood reaches the rete with 10 volume per cent O_2 at P_{O_2} = 0·2 atm and pH = 7·40 before entering the bladder which contains 0·2 atm O_2. While circulating the bladder the blood is acidified to pH 7·0 and P_{O_2} increases to

2 atm (Fig. 10.4B). This causes O_2 to diffuse into the bladder. The blood O_2-content and P_{O_2} will both decrease (Fig. 10.4C). We can assume that there is equilibrium between P_{O_2} in the blood and in the bladder. The P_{O_2} in the blood as it leaves the bladder to enter the rete will depend on the volume of gas in the bladder and upon the blood flow. We assume the P_{O_2} of blood at this point to be 1 atm. It will then contain 7 volume per cent O_2. In the rete O_2 will therefore diffuse from venous to arterial blood raising the arterial P_{O_2} to 1·0 atm and the O_2-content to 12 volume per cent while the venous blood leaving the rete has 5 volume per cent and P_{O_2} = 0·2 atm (Fig. 10.4D). The arterial blood that now enters the bladder will therefore have higher P_{O_2} and higher O_2-content than when it entered the rete (Fig. 10.4E,F). In the bladder it is acidified and P_{O_2} increases to a higher value than the previous 2 atm, say 3 atm. More O_2 diffuses into the bladder and the venous P_{O_2} is also higher than before. Fig. 10.5 shows O_2-equilibrium curves of fish blood and the events shown in Figs. 10.4A to G. The Root-effect is exaggerated.

Thus the P_{O_2} of blood circulating the bladder increases progressively, it is multiplied. This type of concentrating mechanism is termed hairpin counter-current multiplication. The word hairpin refers to the loop nature of the circulation. Steen (1963b) showed that the blood is acidified from pH 7·4 to about 7·0 as it circulates the bladder and Kuhn *et al.* (1963) showed that an acidification with this effect, combined with a rete of certain diffusion characteristics, dimensions and blood flow should be able to produce O_2 pressures of 1000 atm in the bladder.

This mechanism will operate only if the O_2-affinity of arterial and venous blood is maintained while the blood circulates the rete. Since the main parameter affecting O_2-affinity is pH, this point was checked by measuring pH of arterial and venous blood as they entered and left the rete (Steen, 1963b). Surprisingly the pH of venous blood increased, while that of arterial decreased as each passed the rete. This indicated that rete was acid permeable and that the O_2-affinity was changing in rete blood. This observation threw doubt on the validity of the current multiplication theory. A solution to this dilemma, and a dilemma it was since all the other evidence supported the counter-current hypothesis, came from the investigation of Forster and Steen (1969) on the rate of the Root-shift. When eel blood is acidified, as it is in the bladder, O_2 appears in the plasma, i.e. P_{O_2} increases, with a half time of 0·005 sec. However, when the pH of eel blood is increased, as it is in venous blood in the rete, the P_{O_2} decrease has a half time of at least 10 sec. This means that even if the pH of venous blood decreases in venous blood as it circulates the rete, the O_2-affinity does not change markedly during the short time of passage. That such a "delayed reaction" of O_2-affinity to pH changes really was in operation was supported experimentally by Berg and Steen (1968).

REFERENCES

Albers, C. (1961). Der Mechanismus des Wärmehechelns beim Hund. I. Die Ventilation und die arteriellen Blutgase während des Wärmehechelns. *Pflügers Arch. ges. Physiol.* **274,** 125–147.

Andersen, H. T. (1966). Physiological adaptions in diving vertebrates. *Phys. Rev.* **46,** 212–243.

Andersen, H. T. and Løvø, A. (1967). Indirect estimation of partial pressure of oxygen in arterial blood of diving ducks. *Resp. Physiol.* **2,** 163–167.

Anthony, E. H. (1961). Survival of goldfish in presence of carbon monoxide. *J. exp. Biol.* **38,** 109–125.

Atkins, D. (1936). On the ciliary mechanism and interrelationship of Lamellibranches. Part I. New observations on sorting mechanisms. *Q. Jl microsc. Sci.* **79,** 181–309.

Barcroft, J. and Barron, D. H. (1946). Observations upon the form and relations of the maternal and fetal vessels in the placenta of the sheep. *Anat. Rec.* **94,** 569–595.

Barron, D. H. and Battaglia, F. C. (1955–56). The oxygen concentration gradient between the plasmas in the maternal and fetal capillaries of the placenta of the rabbit. *Yale J. Biol. Med.* **28,** 197–207.

Bartels, H. and Metcalfe, J. (1965). Some aspects of the comparative physiology of placental gas exchange. *Int. union. physiol. sci.* **4,** 34–52.

Bartels, H., Hiller, G. and Reinhardt, W. (1966). Oxygen affinity of chicken blood before and after hatching. *Resp. Physiol.* **1,** 345–356.

Baumgarten, D., Randall, D. J., and Malyusse, M. (1968). Beziehungen zwischen Gasaustausch und Ionenaustausch in den Kiemen von Knochenfischen. *Pflügers Arch. ges. Physiol.* **300,** 16.

Belkin, D. A. (1962). Anaerobiosis in diving turtles. *The Physiologist* **5,** 105.

Belkin, D. A. (1963). Diving bradycardia in the iguana. *The Physiologist* **6,** 136.

Belkin, D. A. (1968). Aquatic respiration and underwater survival of two fresh water turtle species. *Resp. Physiol.* **4,** 1–14.

Benesch, R., Benesch, R. E., and Yu, C. I. (1968). Reciprocal binding of oxygen and diphosphoglycerate by human hemaglobin. *Proc. natn. Acad. Sci., U.S.A.* **59,** 526–531.

Berg, T. and Steen, J. B. (1965). Physiological mechanisms for aerial respiration in the eel. *Comp. Biochem. Physiol.* **15,** 469–484.

Berg, T. and Steen, J. B. (1968). The mechanism of oxygen concentration in the swimbladder of the eel. *J. Physiol.* **195,** 631–638.

Berger, R. and Libby W. F. (1969). Equilibration of atmospheric carbon dioxide with sea water: possible enzymatic control of the rate. *Science, N.Y.* **164,** 1395–1397.

Berger, M., Roy, O.Z. and Hart, J. S. (1970). The co-ordination between respiration and wing beats in birds. *Z. vergl. Physiol.* **66,** 190–200.

Biscoe, T. J. and Purves, M. J. (1967). Factors affecting the cat carotid chemoreceptor and cervical sympathetic activity with special reference to passive hind-limb movements. *J. Physiol.* **190,** 425–441.

Black, E. C. (1940). The transport of oxygen by the blood of fresh-water fish. *Biol. Bull.* **79,** 215–229.

Black, E. C., Fry, F. E. J. and Black, V. S. (1954). The influence of carbon dioxide on the utilization of oxygen by some fresh-water fish. *Can. J. Zool.* **32,** 408–420.

Blank, M. and Roughton, F. J. W. (1960). The permeability of monolayers to carbon dioxide. *Trans. Faraday Soc.* **56,** 1832–1841.

Bohr, C., Hasselbalch, K. and Krogh, A. (1904). Ueber einen in biologisches Beziehung wichtigen Einfluss, den die Kohlensäurespannung des Blutes auf dessen Sauerstoff bindung übt. *Skand. Arch. Physiol.* **16,** 402–412.

Bond, A. N. (1960). An analysis of the response of salamander gills to changes in the oxygen concentration of the medium. *Devl. Biol.* **2,** 1–20.

Bretz, W. L. and Schmidt-Nielsen, K. (1970) Patterns of air flow in the duck lung. *Fedn Proc. Fedn Am. Socs. exp. Biol.* 2338.

Buck, J. (1962). Some physical aspects of insect respiration. *A. Rev. Ent.* **7,** 27–56.

Bugge, J. (1961). The heart of the African lung fish *Protopterus. Vidensk. Meddr. dansk. naturh. Foren.* **123,** 193–210.

Burton, R. R. and Smith, A. H. (1969). Induction of cardiac hypertrophy and polycythemia in the developing chick at high altitude. *Fedn Proc. Fedn Am. Socs exp. Biol.* **28** (3), 1170–1177.

Carey, F. G. and Teal, J. M. (1966). Heat conservation in tuna fish muscle. *Proc. natn. Acad. Sci. U.S.A.* **56,** 1464–1469.

Chaetum, E. P. (1934). Limnological investigations on respiration, annual migratory cycle, and other related phenomena in fresh-water pulmonate snails. *Trans. Am. Microsc. Soc.* **53,** 348–407.

Clements, J. A. and Tierney, D. F. (1964). Alveolar instability associated with altered surface tension. *Handb. Physiol. Sec.* 3, **2,** 1565–1585,

Clements, J. A., Hustead, R. F., Johnson, R. P. and Gribetz, I. (1961). Pulmonary surface tensions and alveolar stability. *J. appl. Physiol.* **16,** 444–450.

Cosgrove, W. B. and Schwartz, J. B. (1965). The properties and function of the blood pigment of the earthworm *Lumbricus terrestris. Physiol. Zool.* **38,** 206–212.

Craw, M. R., Constantine, H. P., Morello, J. A. and Forster, R. E. (1963). Rate of the Bohr shift in human red cell suspensions. *J. appl. Physiol.* **18,** 317–324.

Crisp, D. J. (1964). Plastron Breathing. *In*: "Recent Progress in Surface Science," Vol. 2 (J. F. Danielli, K. G. A. Pankhurst and A. C. Riddiford, eds). Academic Press, New York, 377–425.

Czopek, J. (1966). Quantitative studies on the morphology of respiratory surfaces in amphibians. *Acta anat.* **62** (2), 296–323.

Dahr, E. (1924). Die Atmungsbewegungen der Landpulmonaten. *Lunds Univ. Aarsskr., N.F. Avd. 2,* **20,** Nr. 10: 1–19.

Dahr, E. (1927). Studien über die Respiration der Landpulmonaten. *Lunds Univ. Aarsskr. N.F. Avd 2,* **23,** Nr. 10: 1–118.

Dam, L. van (1938). On the utilization of oxygen and regulation of breathing in some aquatic animals. *Dissertation, Groningen.*

Dam, L. van (1954). On the respiration in scallops (Lamellibranchiata). *Biol. Bull.* **107,** 192–202.

DeLong, K. T. (1962). Quantitative analysis of blood circulation through the frog heart. *Science, N.Y.* **138,** 693–694.

Dolk, H. E. and Postma, N. (1927). Ueber die Haut- und die Lungenatmung von *Rana temporaria. Z. vergl. Physiol.* **5,** 417–444.

Drastich, L. (1927). Über das Leben der Salamander-larven bei hohem und nedrigen Sauerstoffpartialdruch. *Z. vergl. Physiol.* **2,** 632–657.

Ege, R. (1918). On the respiratory function of the air stores carried by some aquatic insects (*Corixidae*, *Dytiscidae* and *Notonecta*). *Z. allg. Physiol.* **17,** 81–124.

Enns, T., Scholander, P. F. and Bradstreet, E. D. (1965). Effect of hydrostatic pressure on gases dissolved in water. *J. physical Chem., Ithaca* **69,** 389–391.

Faber, J. J. and Hart, F. M. (1966). The rabbit placenta as an organ of diffusional exchange. *Circulation Res.* 816–833.

Fish, G. R. (1956). Some aspects of the respiration of six species of fish from Uganda. *J. exp. Biol.* **33,** 186–195.

Forster, R. E. and Steen, J. B. (1968). Rate limiting processes in the Bohr shift in human red cells. *J. Physiol.* **196,** 541–562.

Forster, R. E. and Steen, J. B. (1969). The rate of the "Root shift" in eel red cells and eel haemoglobin solutions. *J. Physiol.* **204,** 259–282.

Fox, C. J. J. (1909). On the coefficients of absorption of nitrogen and oxygen in distilled water and sea water, and of atmospheric carbonic acid in sea water. *Trans. Faraday Soc.* **5,** 68–87.

Fox, H. M. (1945). The oxygen affinities of certain invertebrate haemoglobins. *J. exp. Biol.* **21,** 161–165.

Fox, H. M. (1955). The effect of oxygen on the concentration of haem in invertebrates. *Proc. R. Soc.* **B 143,** 203–214.

Foxon, G. E. H. (1964). "Blood and Respiration." *In*: "Physiology of the Amphibia" (J. A. Moore, ed.). Academic Press, New York.

Fraenkel, G. (1930). Der Atmungsmechanismus des Skorpions. *Z. vergl. Physiol.* **11,** 656–661.

Ghiretti, F. (1966a). Respiration. *In*: "Physiology of Molluscs." (K. M. Wilbur and C. M. Yonge, eds.) Academic Press, New York.

Ghiretti, F. (1966b). Molluscan hemocyanins. In: "Physiology of Molluscs" (K. M. Wilbur and C. M. Yonge, eds). Academic Press, New York.

Goode, R. C., Brown, E. B. jr., Howson, M. G. and Cunningham, D. J. C. (1969). Respiratory effects of breathing down a tube. *Resp. Physiol.* **6,** 343–360.

Goodwin, T. W. (1960). Biochemistry of pigments. *In*: "Physiology of Crustacea", Vol. 1 (T. H. Waterman, ed.). Academic Press, New York.

Gray, I. E. (1954). Comparative study of the gill area of marine fishes. *Biol. Bull.* **107,** 219–225.

Gray, I. E. (1957). A comparative study of the gill area of crabs. *Biol. Bull.* **112,** 34–42.

Grover, R. F. (1965). Pulmonary circulation in animals and man at high altitude. *Ann. N.Y. Acad. Sci.* **127,** 632–639.

Hack, H. R. B. (1956). An application of a method of gas micro-analysis to the study of soil air. *Soil Sci.* **82,** 217–232.

Hall, F. G. (1930). The ability of the common mackerel and certain other marine fishes to remove dissolved oxygen from the water. *Am. J. Physiol.* **93,** 417–421.

Hall, F. G. and Gray, I. E. (1929). The hemoglobin concentration of the blood of marine fishes. *J. biol. Chem.* **81,** 589–594.

Hall, F. G., Dill, D. B. and Barron, E. S. G. (1936). Comparative physiology in high altitudes. *J. cell. comp. Physiol.* **8,** 301–313.

Hargitay, B. and Kuhn, W. (1951). Das Multiplikationsprinsip als Grundlage der Harnkonzentrierung in der Niere. *Z. Elektrochem. angew. phys. Chem.* **55,** 539–558.

Hart, J. S. and Roy, O. Z. (1966). Respiratory and cardiac responses to flight in pigeons. *Physiol. Zoöl.* **39,** 291–306.

Harvey, E. (1928). The oxygen consumption of luminous bacteria. *J. Gen. Physiol.* **11,** 232–242.

Haughton, T. M., Kerkut, G. A. and Munday, K. A. (1958). The oxygen dissociation and alkaline denaturation of haemoglobin from two species of earthworms. *J. exp. Biol.* **35,** 360–368.

Hays, F. A. and Sumbardo, A. H. (1927). Physical characters of eggs in relation to hatchability. *Poult. Sci.* **6,** 196–200.

Hazelhoff, E. H. (1926). "Regeling der Adembaling by Insecten en Spinnen." Proefschrift, Utrecht.

Hazelhoff, E. H. (1939). Über die Ausnutzung des Sauerstoffs bei verschiedenen Wassertieren. *Z. vergl. Physiol.* **26,** 306–327.

Hazelhoff, E. H. (1951). Structure and function of the lung of birds. *Poult. Sci.* **30,** 3–10.

Heller, J. (1930). Sauerstoffverbrauch der Schmetterlingspuppen in Abhängigkeit von der Temperatur. *Z. vergl. Physiol.* **11,** 448–460.

Hemingway, A. and Barbour, H. H. (1938). The thermal tolerance of normal resting dogs as measured by changes in the acid-base equilibrium and the dilution-concentration effect of plasma. *Am. J. Physiol.* **124,** 164–270.

Hemmingsen, E. A. (1965). Transfer of oxygen through solutions of heme pigments. *Acta physiol. scand., Suppl.* **246,** 53 pp.

Hemmingsen, E. A. and Douglas, E. L. (1970). Respiratory characteristics of the hemoglobin-free fish *Chaenocephalus aceratus. Comp. Biochem. Physiol.* **32,** 733–744.

Hinton, H. E. (1953). Some adaptions of insects to environments that are alternately dry and flooded, with some notes on the habits of the Stratiomyidae. *Trans. Soc. Br. Ent.* **11,** 209–227.

Holland, R. A. B. and Forster, R. E. (1966). The effect of size of red cells on the kinetics of their oxygen uptake. *J. gen. Physiol.* **49,** 727–742.

Hughes, G. M. (1960). A comparative study of gill ventilation in marine teleosts. *J. exp. Biol.* **37,** 28–45.

Hughes, G. M. (1961). How a fish extracts oxygen from water. *New Scientist* **11,** 346–348.

Hughes, G. M. (1966). The dimensions of fish gills in relation to their function. *J. exp. Biol.* **45,** 177–195.

Hughes, G. M. and Grimstone, A. V. (1965). The fine structure of the secondary lamellae of the gills of *Gadus pollachius. Q. Jl microsc. Sci.* **106,** 343–353.

Hughes, G. M., Knights, B. and Scammell, C. A. (1969). The distribution of P_{O_2} and hydrostatic pressure changes within the branchial chamber in relation to gill ventilation of the shore crab *Carcinus maenas. J. exp. Biol.* **50,** 1–17.

Hurtado, A. (1964). Animals in high altitudes: resident man. *In*: "Handbook of Physiology." Sec. 4, Adaption to the environment (D. B. Dill, ed.). American Physiology Society, Washington D.C., 843–860.

Hyman, L. H. (1955). "The Invertebrates." Vol. IV. Echinodermata, Water vascular system. McGraw-Hill, London. 468–473.

Irving, L. (1939). Respiration of diving mammals. *Phys. Rev.* **19,** 112–134.

Irving, L., Scholander, P. F. and Grinnell, S. W. (1942). The regulation of arterial blood pressure in the seal during diving. *Am. J. Physiol.* **135,** 557–566.

Ito, T. (1953). The permeability of the integument to oxygen and carbon dioxide *in vivo. Biol. Bull.* **105,** 308–315.

Jakubowski, M., Byczkowska-Smyk, W. and Mikhalev, Y. (1969). Vascularization and size of the respiratory surfaces in the antarctic white-blooded fish *Chaenichthys rugosus* Regan (Percoidei, Chaenichthydea). *Zool. Poloniae* **19,** 303–317.

Johansen, K. (1964). Regional distribution of circulating blood during submersion asphyxia in the duck. *Acta physiol. scand.* **62,** 1–9.

Johansen, K. (1965). Cardiac output in the large cephalopod *Octopus dofleini*. *J. exp. Biol.* **42,** 475–480.

Johansen, K. (1966). Airbreathing in the teleost *Symbranchus mamoratus*. *Comp. Biochem. Physiol.* **18,** 383–395.

Johansen, K. (1968). Air-breathing fishes. *Scient. Am.* **219,** 102–111.

Johansen, K. (1970). Airbreathing in fishes. *In*: "Fish Physiology" Vol. 4 (W. S. Hoar and D. J. Randell, eds). Academic Press, New York.

Johansen, K. and Martin, A. W. (1962). Circulation in the cephalopod *Octopus dofleini*. *Comp. Biochem. Physiol.* **5,** 161–176.

Johansen, K. and Lenfant, C. (1966). Gas exchange in the cephalopod, *Octopus dofleini*. *Am. J. Physiol.* **210,** 910–918.

Johansen, K. and Martin, A. W. (1966). Circulation in a giant earthworm, *Glossoscolex giganteus*. II. Resp. prop. of the blood and some patterns of gas exchange. *J. exp. Biol.* **45,** 165–172.

Johansen, K. and Lenfant, C. (1967). Respiratory function in the South America lungfish *Lepidosiren paradoxa* (Fitz.) *J. exp. Biol.* **46,** 205–218.

Johansen, K. and Lenfant, C. (1968). Respiration in the African lungfish *Protopterus aethiopicus*. II. Control of breathing. *J. exp. Biol.* **49,** 453–468.

Johansen, K. and Hanson, D. (1968). Functional anatomy of the hearts of lungfishes and amphibians. *Am. Zoologist*, **8,** 191–210.

Johansen, K., Lenfant, C. and Grigg, G. C. (1966). Respiratory properties of blood and responses to diving of the platypus, *Ornithorhynchus anatinus* (Shaw). *Comp. Biochem. Physiol.* **15,** 597–608.

Johansen, K., Lenfant, C., and Hanson, D. (1968). Cardiovascular dynamics in the lungfishes. *Z. vergl. Physiol.* **59,** 157–186.

Johansen, K., Lenfant, C., Schmidt-Nielsen, K. and Petersen, J. A. (1968). Gas exchange and control of breathing in the electric eel, *Electrophorus electricus*. 2 *vergl. Physiol.* **61,** 137–163.

Johansen, K., Hanson, D. and Lenfant, C. (1970). Respiration in a primitive air breather, *Amia calva*. *Resp. Physiol.* **9,** 162–175.

Johansen, K., Lenfant, C. and Mecklenburg, T. A. In press. Respiration in the crab, *Cancer magister*. *Resp. Physiol.*

Johnson, M. L. (1941/2). The respiratory function of the haemoglobin of the earthworm. *J. exp. Biol.* **18,** 266–277.

Kanwisher, J. W. (1966). Tracheal gas dynamics in pupae of the *Cecropia* Silkworm. *Biol. Bull.* **130,** 96–105.

Kao, F. (1963). An experimental study of the pathways involved in exercise hyperpnoea employing cross-circulation techniques. *In*: "The Regulation of Human Respiration" (D. J. C. Cunningham and B. B. Lloyd, eds). Blackwell, Oxford.

Kellogg, R. H. (1968). Altitude acclimatization. A historical introduction emphasizing the regulation of breathing. *The Physiologist*, **11,** 37–57.

Klika, E. and Lelek, A. (1967). A contribution to the study of the lungs of the *Protopterus annectens* and *Polypterus senegalensis*. *Folia Morphologica*, **15,** 168–175.

Krogh, A. (1904). On cutaneous and pulmonary respiration of the frog. *Skand. Arch. Physiol.* **15,** 328–419.

Krogh, A. (1910). On the mechanism of the gas-exchange in the lungs. *Skand. Arch. Physiol.* **23,** 248–278.

Krogh, A. (1920). Studien über Tracheen respiration. II. Über Gasdiffusion in der Tracheen. III. Die Kombination von mechanischer Ventilation mit Gasdiffusion nach Versuchen an Dyctiluslarven. *Pflügers Arch. ges. Physiol.* **179,** 95–112, 113–120.

Krogh, A. (1922). The anatomy and physiology of capillaries. Yale University Press. Reprinted 1959 (Hafner Publishing Co. New York).

Krogh, A. (1941). The comparative physiology of respiratory mechanisms. University Pennsylvania Press, Philadelphia. 172 pp.

Kuhn, W. and Kuhn, H. J. (1961). Multiplication von Aussalz- und Einzeleffecten für die Bereitung hoher Gasdrücke in der Schwimmblase. *J. elektrochem. Soc.* **65,** 426–439.

Kuhn, W., Ramel, A. Kuhn., H. J. and Marti, E. (1963). The filling mechanism of the swimbladder. *Experientia*, **19,** 497–511.

Kutchai, H. and Steen, J. B. Permeability of the shell and shell membranes of hens' eggs during development. *Resp. Physiol.* In press.

Larimer, J. J. and Schmidt-Nielsen, K. (1960). A comparison of blood carbonic anhydrase of various animals. *Comp. Biochem. Physiol.* **1,** 19–23.

Laverack, M. S. (1963). The physiology of earthworms. *Int. ser. mon. Pure. Appl. Biol.* **15.**

LeFebvre, E. (1964). The use of D_2O^{18} for measuring the energy metabolism in *Columbia livia*, at rest and in flight. *Auk.* **81,** 403–416.

Lenfant, C. (1969), Physiological properties of blood of marine mammals. *In*: "The Biology of Marine Mammals" (H. T. Andersen, ed.). Academic Press, New York.

Lenfant, C. and Johansen, K. (1965). Gas transport by hemocyanin-containing blood of the Cephalopod *Octopus dofleini. Am. J. Physiol.* **209,** 991–998.

Lenfant, C. and Johansen, K. (1966). Respiratory function in the elasmobranch *Squalus sucklyi. Resp. Physiol.* **1,** 13–29.

Lenfant, C. and Johansen, K. (1967). Respiratory adaptions in selected amphibians. *Resp. Physiol.* **2,** 247–260.

Lenfant, C. and Johansen, K. (1968). Respiration in the African lungfish *Protopterus aethiopicus*. I. Respiratory properties of blood and normal patterns of breathing and gas exchange. *J. exp. Biol.* **49,** 437–453.

Lenfant, C., Johansen, K. and Grigg, G. C. (1966/7). Respiratory properties of blood and pattern of gas exchange in the lungfish *Neoceratodus forsteri* (Krefft). *Resp. Physiol.* **2,** 1–21.

Lenfant, C., Johansen, K., Petersen, A. J. and Schmidt-Nielsen, K. (1969). Gas exchange in the airbreathing fishes, *Lepidosiren paradoxa* and *Symbranchus marmoratus*. In preparation.

Lenfant, C., Torrance, J., English, E., Finch, C. A., Reynafarje, C., Ramos, J. and Faura J. (1968). Effect of altitude on oxygen binding by hemoglobin and on organic phosphate levels. *J. clin. Invest.* **47,** 2652–2658.

Lenfant, C., Johansen, K., Petersen, J. A. and Schmidt-Nielsen, K. (1970). Respiration in the fesh water turtle. *Chelys fimbriata. Resp. Physiol.* **8,** 261–275.

Levy, R. I. and Schneiderman H. A. (1958). An experimental solution to the paradox of discontinuous respiration in insects. *Nature, Lond.* **182,** 491–493.

Lindroth, A. (1938). Atmungsregulation bei *Astacus fluviatilis Ark. Zool.* **30B** No. 3, 1–7.

Longmuir, I. S. and Bourke, A. (1960). The measurement of the diffusion of oxygen through respiring tissue. *Biochem. J.* **76,** 225–229.

Longo, L. D., Schwarz, R. H. and Forster, R. E. II (1968). O_2 tension of blood from uterine wall venules, maternal placental venules, and uterine vein. *J. appl. Physiol.* **24,** 787–791.

Maas, J. A. (1939). Über die Atmung von *Helix pomatia. Z. vergl. Physiol.* **26,** 605–610.

MacDougall, J. D. B. and McCobe, M. (1967). Diffusion coefficient of oxygen through tissues. *Nature, Lond.* **215,** 1173–1174.

McCutcheon, F. H. (1936). Hemoglobin function during the life history of the bull-frog. *J. cell. comp. Physiol.* **8,** 63–81.

Manwell, C. (1958). A "Fetal-maternal shift" in the ovoviviparous spiny dogfish *Squalus suckleyi* (Girard). *Physiol. Zoöl.* **31,** 93–100.

Manwell, C. (1958). The oxygen-respiratory pigment equilibrium of the Hemocyanine and myoglobin of the amphineuron mollusc *Cryptochiton stelleri. J. cell. comp. Physiol.* **52,** 341–353.

Manwell, C. (1963). The chemistry and biology of hemoglobin in some marine clams. I. Distribution of the pigment and properties of the oxygen equilibrium. *Comp. Biochem. Physiol.* **8,** 209–218.

Maren, T. H. (1967). Carbonic Anhydrase: Chemistry, physiology and distribution. *Phys. Rev.* **47,** 597–781.

Metcalfe, J., Moll, W., Bartels, H., Hilpert, P. and Parer, J. T. (1965). Transfer of carbon monoxide and nitrous oxide in the artificially perfused sheep placenta. *Circulation Res.* **16,** 95–101.

Metcalfe, J., Bartels, H. and Moll, W. (1967). Gas exchange in the pregnant uterus. *Phys. Rev.* **47,** 782–838.

Miller, P. L. (1960). Respiration in the desert locust. III. Ventilation and spiracles during flight. *J. exp. Biol.* **37,** 264–278.

Miller, P. L. (1964). Respiration—Aerial gas transport. *In:* "The Physiology of Insecta", Vol. 3 (M. Rockstein, ed.). Academic Press, New York.

Miller, P. L. (1966). The regulation of breathing in insects. *In*: "Advances in Insect Physiology", Vol. 3 (J. W. L. Beament, J. E. Treherene and V. B. Wigglesworth, eds). Academic Press, London. 279–354.

Mitchell, R. A. (1966). The role of medullary chemoreceptors in acclimation to high altitude. Proc. Int. Symp. Cardiovasc. Respir. Effects Hypoxia. Karger, Basel/ New York.

Moll, W. (1966). Der Progress der gleichzeitigen Diffusion und Reaktion von Hämoglobin und Sauerstoff bei der Sauerstoff-Aufname und -Abgabe des Blutes. Habil. Schrift, Tübingen.

Mossman, H. W. (1926). The rabbit placenta and the problem of placental transmission. *Am. J. Anat.* **37,** 433–497.

Muir, B. S. and Kendall, J. I. (1968). Structural modifications in the gills of tunas and some other oceanic fishes. *Copeia*, **2,** 388–398.

Muir, B. S. and Hughes, G. M. (1969). Gill dimensions for three species of tunny. *J. exp. Biol.* **51,** 271–286.

Mulhausen, R., Astrup, P. and Kjeldsen, K. (1967). Oxygen affinity of hemoglobin in patients with cardiovascular diseases, anemia and cirrhosis of the liver. *Scand. J. Lab. Clin. Med.* **19,** 291–299.

Munshi, D. (1968). The accessory respiration organs of *Anabas testidineus* (Bloch) (Anabandital, Pisces). *Proc. Linn. Soc. Lond.* **179,** 107–126.

Nemenz, H. (1960). Beiträge zur Kenntnis der Biologie von *Ephydra cinerca* (Jones) (Diptera, Ephydridoe). *Zool. Anz.* **165,** 218–226.

Nicloux, M. (1923). Action de l'oxyde de carbone sur les poissons et capacité respiratoire du sang de ces animaux. *C.r. Séanc. Soc. Biol.* **89,** 1328–1331.

Novy, M. J. and Parer, J. T. (1969). Absence of high blood oxygen affinity in the fetal cat. *Resp. Physiol.* **6,** 144–151.

Paganelli, C. V., Bateman, N. and Rahn, H. (1967). Artificial gills for gas exchange in water. *In*: "Proceedings 3. Symposium Underwater Physiology", Ch. 38 (C. J. Lambertsen, ed.). Williams and Wilkins, Baltimore.

Parer, J. T., (1970). Oxygen transport in human subjects with hemoglobin variants having altered oxygen affinity. *Resp. Physiol.* **2,** 43–49.

Parry, G. (1966). Osmotic adaption in fishes. *Biol. Rev.* **41,** 392–444.

Pasztor, V. M. and Kleerekoper, H. (1962). The roles of the gill filament musculature in teleosts. *Can. J. Zool.* **40,** 785–802.

Pattle, R. E. (1966). Surface tension and thc lining of the lung alveoli. *In*: "Advances in Respiratory Physiology" (C. G. Caro, ed.). Arnold, London.

Pattle, R. E. and Hopkinson, D. A. W. (1963). Lung lining in bird, reptile and amphibian. *Nature, Lond.* **200,** 894–902.

Pearson, O. P. (1964). Metabolism and heat loss during flight in pigeons. *Condor,* **66,** 182–185.

Pelseneer, P. (1935). "Essay d'ethologie zoologique". 662 pp. Acad. Roy. Belg. Classe. Sci. Publ. Fond. Aqathon Potter, Brussels.

Perútz, M. F. (1969) The haemoglobin molecule. The Croonian lecture 1968. *Proc. Roy. Soc. B.* **173,** 113–140.

Piiper, J. and Schumann, D. (1967). Efficiency of O_2 exchange in the gills of the dogfish, *Scyliorhinus stellaris*. *Resp. Physiol.* **2,** 135–148.

Piiper, J. and Baumgarten-Schumann, D. (1968). Effectiveness of O_2 and CO_2 exchange in the gills of the dogfish (*Scyliorhinus stellaris*). *Resp. Physiol.* **5,** 338–349.

Pilson, M. E. (1965). Variation of hemocyanin concentration in the blood of four species of haliotis. *Biol. Bull.* **128,** 459–472.

Power, G. G., Longo, L. D., Wagner, H. N. Jr., Kuhl, D. K. and Forster, R. E. II (1967). Uneven distribution of maternal and fetal placental blood flow, as demonstrated using macroaggregates, and its response to hypoxia. *Clin. Invest.* **46,** 2053–2063.

Precht, H. (1939). Die Lungeatmung der Süsswasserpulmonaten. *Z. vergl. Physiol.* **26,** 696–738.

Prosser, C. L. and Brown, F. (1962). "Comparative Animal Physiology," 2nd ed. W. B. Saunders Co., London.

Rahn, H. and Paganelli, C. V. (1968). Gas exchange in gas gills of diving insects. *Resp. Physiol.* **5,** 145–164.

Rakestraw, N. W. and Emmel, E. W. M. (1938). The solubility of nitrogen and argon in sea water. *J. Phys. Chem.* **42,** 1211–1215.

Read, K. R. H. (1966). Molluscan hemoglobin and myoglobin. *In*: "Physiology of Mollusca" (K. M. Wilbur and C. M. Yonge, eds). Academic Press, New York.

Redmond, J. R. (1955). The respiratory function of hemocyanin in Crustacea. *J. cell. comp. Physiol.* **46,** 209–247.

Redmond, J. R. (1962). The respiratory characteristics of chiton hemocyanins. *Physiol. Zoöl.* **35,** 304–313.

Remotti, E. (1933). Development of allantoic circulation in response to external variations in gas content. *Bull. Mus. Lab. Zool. Anat. Comp. Univ. Genova* **13,** 1–19.

Richards, A. B. (1957). Studies on arthropod cuticle. XIII. The penetration of dissolved oxygen and electrolytes in relation to the multiple barriers of the epicuticle. *J. Insect Physiol.* **1,** 23–29.

Riess, J. A. (1881). Der Bau der Keimenblatter bei Knochenfischen. *Arch. Naturgesch.* **47,** I: 518–550.

Riggs, A. (1959–60). The nature and significance of the Bohr effect in mammalian hemoglobin. *J. gen. Physiol.* **43,** 737–752.

Robinson, D. (1939). The muscle hemoglobin of seals as an oxygen store in diving. *Science, N.Y.* **90,** 267–277.

Romanoff, A. L. and Romanoff, A. J. (1949). "The Avian Egg." Wiley, New York.

Romijn, C. and Roos, J. (1938). The air space of the hen's egg and its changes during the period of incubation. *J. Physiol.* **94,** 365–379.

Romijn, C. and Lokhorst, W. (1960). Foetal heat production in the fowl. *J. Physiol.* **150,** 239–249.

Root, R. W. (1931). The respiratory function of the blood of marine fishes. *Biol. Bull.* **61,** 427–456.

Root, R. W. and Irving, L. (1943). The effect of carbon dioxide and lactic acid on the oxygen-combining power of whole and hemolyzed blood of the marine fish, *Tautoga onitis* (Linn.). *Biol. Bull.* **88,** 207–212.

Rossi-Fanelli, A. and Antonini, E. (1957). A new type of myoglobin isolated and crystallized from the muscles of Aplysiae. *Biochemistry* (*USSR*) (*English Transl.*) **22,** 312–321.

Roughton, F. J. W. (1963). Kinetics of gas transport in the blood. *Br. med. Bull.* **19,** 80–89.

Ruud, J. T. (1954). Vertebrates without erythrocytes and blood pigments. *Nature, Lond.* **173,** 848–853.

Salt, G. W. and Zeuthen, E. (1960). The respiratory system. *In*: "Biology and Comparative Physiology of Birds" (A. J. Marshall, ed.). Academic Press, New York.

Satchell, G. H. (1960). The reflex co-ordination of the heart beat with respiration in the dogfish. *J. exp. Biol.* **37,** 719–731.

Sato, T. (1931). Untersuchungen am Blut der gemeinen japanischen Arche-muschel (*Arca inflata*). *Z. vergl. Physiol.* **14,** 763–783.

Saunders, R. L. (1962). The irrigation of the gills in fishes. II. Efficiency of oxygen uptake in relation to respiratory flow, activity and concentration of oxygen and carbon dioxide. *Can. J. Zool.* **40,** 817–862.

Savage, R. M. (1935). The ecology of young tadpoles, with special reference to some adaptions to the habitat of mass spawning in *Rana temporaria*. *Linn. Proc. Zool. Soc.* (*Lond.*), 605–610.

Schmidt-Nielsen, K. (1964). Desert animals, physiological problems of heat and water. Oxford Univ. Press, London.

Schmidt-Nielsen, K. and Larimer, J. J. (1958). Oxygen dissociation curves of mammalian blood in relation to body size. *Am. J. Physiol.* **195,** 424–428.

Schmidt-Nielsen, K., Kanwisher, J., Lasiewski, R. C., Cohn, J. E. and Bretz, W. L. (1969). Temperature regulation and respiration in the ostrich. *The Condor*, **71,** 341–352.

Schneiderman, H. A. and Williams, C. H. (1955). An experimental analysis of the discontinuous respiration of the *Cecropia* silkworm. *Biol. Bull.* **109,** 123–145.

Scholander, P. F. (1940). Experimental investigations on the respiratory function of diving mammals and birds. *Hvalråd. Skr.* **22,** 1–131.

Scholander, P. F. (1954). Secretion of gases against high pressures in the swimbladder of deep sea fishes. II. The *rete mirabile*. *Biol. Bull.* **107,** 260–277.

Scholander, P. F. (1960). Oxygen transport through hemoglobin solutions. How does the presence of hemoglobin in a wet membrane mediate an eightfold increase in oxygen passage? *Science, N.Y.* **131,** 585–590.

Scholander, P. F. and Dam, L. van (1954). Secretion of gases against high pressures in the swimbladder of deep sea fishes. I. Oxygen dissociation in blood. *Biol. Bull.* **107,** 247–259.

Schultz, H. (1958). Die submikroskopische Morphologie des Kiemenepitels. *Int. Congr. Electr. Microsc.* **4,** 421–426.

Sejrsen, P. (1968). Epidermal diffusion barrier to ^{133}Xe in man and studies of clearance of ^{133}Xe by sweat. *J. appl. Phys.* **24,** 211–216.

Semper, C. (1878). Über die Lunge von *Birgus latro*. *Z. wiss. Zool.* **30,** 282–287.

Sørensen, S. C. and Servinghaus, J. W. (1968). Irreversible respiratory insensitivity to acute hypoxia in man born at high altitude. *J. appl. Phys.* **25,** 217–220.

Staub, N. C. (1963). The independence of pulmonary structure and function. *Anaesthesiology* **24,** 831–854.

Staub, N. C. (1969). Respiration. *A. Rev. Physiol.* **31,** 173–203.

Steen, J. B. (1963a). The physiology of the swimbladder of the eel, *Anguilla vulgaris*. I. The solubility of gases and the buffer capacity of blood. *Acta physiol. scand.* **58,** 124–137.

Steen, J. B. (1963b). The physiology of the swimbladder in the eel *Anguilla vulgaris*. III. The mechanism of gas secretion. *Acta physiol. scand.* **59,** 221–241.

Steen, J. B. (1965). Comparative aspects of the respiratory gas exchange of sea urchins. *Acta physiol. scand.* **63,** 164–170.

Steen, J. B. and Berg, T. (1966). The gills of two species of haemoglobin-free fishes compared to those of other teleosts, with a note on severe anaemia in an eel. *Comp. biochem. physiol.* **18,** 517–526.

Steen, J. B. and Kruysse, A. (1964). The respiratory function of teleostan gills. *Comp. Biochem. Physiol.* **12,** 127–142.

Steen, J. B. and Turitzin, S. N. (1968). The nature and biological significance of the pH difference across red cell membranes. *Resp. Physiol.* **5,** 234–242.

Strawinski, S. (1956). Vascularization of respiratory surfaces in ontogeny of the edible frog, *Rana esculenta*. *L. Zool. Polon.* **7,** 327–365.

Stride, G. O. (1958). The application of a Bernoulli equation to problems of insect respiration. *Proc. 10th Intern. Congr. Entomol.* **2,** 335–336.

Sturkie, P. D. (1954). "Avian Physiology." Comstock Publishing Co. Ithaca, New York.

Sutnick, A. I. and Soloff, L. A. (1962). Surface tension reducing activity in the normal and atelectatic human lung. *Am. J. Med.* **35,** 31–50.

Tenney, S. M. and Remmers, J. E. (1963). Comparative quantitative morphology of the mammalian lung: diffusing area. *Nature, Lond.* **197,** 54–57.

Tuft, P. H. (1950). The structure of the insect egg-shell in relation to the respiration of the embryo. *J. exp. Biol.* **26,** 327–334.

Valdivia, E. (1956). Mechanisms of natural acclimation. Capillary studies at high altitudes. Rept. 55–101. School Aviation Med. USAF Randolph Field, Texas.

Van Liere, E. J. and Stickney, J. C. (1963). "Hypoxia". Chicago University Press, Chicago.

Visschedijk, A. H. J. (1968a). The air space and embryonic respiration. 1. The pattern of gaseous exchange in the fertile egg during the closing stages of incubation. Reprinted from *Br. Poult. Sci.* **9,** 173–184.

Visschedijk, A. H. J. (1968b). The air space and embryonic respiration. 2. The times of pipping and hatching as influenced by an artificially changed permeability of the shell over the air space. *Br. Poult. Sci.* **9,** 185–196.

Visschedijk, A. H. J. (1968c). The air space and embryonic respiration. 3. The balance between oxygen and carbon dioxide in the air space of the incubating chicken egg and its role in stimulating pipping. *Br. Poult. Sci.* **9,** 197–209.

Wallengren, H. (1914). Physiolog.—Biolog. Studien über die Atmung bei den Arthropoden. III. Die Atmung der Aeschnalarven. *Lunds Univ. Aarsskr. N.F. Avd.* 2, **10,** No. 8: 1–28.

Wangenstein, O. D., Wilson, D. and Rahn, H. (1969). Diffusive permeability of egg shell to gases. *The Physiologist*, **12,** 385.

Weibel, E. R. (1963). "Morphometry of the Human Lung." Springer Verlag, Berlin.

Weis-Fogh, T. (1961). Power in flapping flight. *In*: "The Cell and the Organism" (J. A. Ramsay and V. B. Wigglesworth, eds). Cambridge University Press. 283–300.

Weis-Fogh, T. (1964a). Functional design of the tracheal system of flying insects as compared with the avian lung. *J. exp. Biol.* **41,** 207–228.

Weis-Fogh, T. (1964b). Diffusion in insect wing muscle, the most active tissue known. *J. exp. Biol.* **41,** 229–256.

Weis-Fogh, T. (1964c). Biology and physics of locust flight. VIII. Lift and metabolic rate of flying locusts. *J. exp. Biol.* **41,** 257–272.

White, P. T. (1968). Experimental studies on the heart of the chick embryo at two stages of development (Ph.D. thesis, 1968). University of London Library, Senate House, London W.C.1.

Whitford, W. G. and Hutchinson, W. H. (1965). Gas exchange in the salamanders. *Physiol. Zoöl.* **33,** 228–242.

Wigglesworth, V. B. (1930). A theory of tracheal respiration in insects. *Proc. R. Soc.* **B106,** 230–250.

Wigglesworth, V. B. (1953). Surface forces in the tracheal system of insects. *Q. Jl Microsc. Sci.* **94,** 507–522.

Willmer, E. N. (1934). Some observations on the respiration of certain tropical fresh-water fishes. *J. exp. Biol.* **11,** 283–306.

Winterstein, H. (1908). Beiträge zur Kenntnis der Fischatmung. *Pflügers Arch. ges. Physiol.* **125,** 73–98.

Wolvekamp, H. P. and Waterman, T. H. (1960). Respiration. *In*: "The Physiology of Crustacea", Vol. 1 (T. H. Waterman, ed.). Academic Press, New York.

Wolvekamp, H. P., Baerends, G. P, Kok, B. and Mommaerts, W. F. H. M. (1942). O_2 and CO_2 binding properties of the blood of the catfish (*Sepia officinalis*) and the common squid (*Loligo vulgaris*). *Arch. néerl. physiol.* **26,** 203–218.

Yonge, C. N. (1947). The pallid organs in the aspidobranch Gastropoda and their evolution throughout the Mollusca. *Phil. Trans. R. Soc.* **B222,** 443–518.

Zeuthen, E. (1942). The ventilation of the respiratory tract in birds. *KK. danske Vidensk. Selsk, Biol. Med.* **17,** 1–50.

AUTHOR INDEX

Numbers in italics are those pages on which references are listed in full

A

Albers, C., 104, *162*
Andersen, H. T., 132, 133, 134, 135, 138, *162*
Anthony, E. H., 54, *162*
Antonini, E., 28, *170*
Astrup, P., 126, *168*
Atkins, D., 26, *162*

B

Baerends, G. P., 27, *172*
Barbour, H. H., 104, *165*
Barcroft, J., 139, *162*
Barron, D. H., 139, *162*
Barron, E. S. G., 103, 131, *164*
Bartels, H., 139, 140, 141, 142, 143, 144, 152, *162*, *168*
Bateman, N., 93, *168*
Battaglia, F. C., 139, *162*
Baumgarten, D., 55, *162*
Baumgarten-Schumann, D., 46, *169*
Belkin, D. A., 132, 135, 138, *162*
Benesch, R., 11, 130, *162*
Berg, T., 26, 37, 48, 50, 54, 58, 159, 161, *162*, *171*
Berger, R., 17, 121, *162*
Biscoe, T. J., 105, *162*
Black, E. C., 51, 52, *163*
Black, V. S., 51, 52, *163*
Blank, M., 81, *163*
Bohr, C., 8, 9, *163*
Bond, A. N., 76, *163*
Bourke, A., 5, *167*
Bradstreet, E. D., 16, *164*
Bretz, W. L., 121, *163*, *170*
Brown, E. B. Jr., 106, *164*
Brown, F., 21, 99, 100, 142, *169*
Buck, J., 82, 83, 87, *163*
Bugge, J., 66, *163*
Burton, R. R., 126, *163*
Byczkowska-Smyk, W., 54, *165*

C

Carey, F. G., 20, *163*
Chaetum, E. P., 80, *163*
Clements, J. A., 103, *163*
Cohn, J. E., 121, *170*
Constantine, H. P., 14, *163*
Cosgrove, W. B., 79, *163*
Craw, M. R., 14, *163*
Crisp, D. J., 92, 93, 94, *163*
Cunningham, D. J. C., 106, *164*
Czopek, J., 80, 100, *163*

D

Dahr, E., 97, *163*
Dam, L. van, 45, 46, 158, 160, *163*, *170*
DeLong, K. T., 70, *163*
Dill, D. B., 130, 131, *164*
Dolk, H. E., 80, *163*
Douglas, E. L., 54, 55, *165*
Drastich, L., 76, *163*

E

Ege, R., 92, 93, *164*
Emmel, E. W. M., 17, *169*
English, E., 129, *167*
Enns, T., 16, *164*

F

Faber, J. J., 139, *164*
Faura, J., 129, *167*
Finch, C. A., 129, *167*
Fish, G. R., 49, 51, *164*
Forster, R. E., 13, 14, 15, 53, 145, 161, *163*, *164*, *165*, *167*
Forster, R. E. II, 141, 143, *169*
Fox, C. J. J., 17, *164*
Fox, H. M., 28, 32, *164*
Foxon, G. E. H., 75, 104, *164*
Fraenkel, G., 96, *164*
Fry, F. E. J., 51, 52, *163*

G

Ghiretti, F., 26, 97, *164*
Goode, R. C., 106, *164*
Goodwin, T. W., 32, *164*
Gray, I. E., 26, 31, 37, 48, *164*
Gribetz, I., 103, *163*
Grigg, G. C., 12, 67, 73, 134, *166*, *167*
Grimstone, A. V., 37, 38, *165*
Grinnell, S. W., 135, *165*
Grover, R. F., 130, *164*

H

Hack, H. R. B., 77, *164*
Hall, F. G., 48, 52, 130, 131, *164*
Hanson, D., 33, 60, 64, 66, 68, 69, 70, 71, 72, 73, *166*
Hargitay, B., 158, *164*
Hart, F. M., 139, *164*
Hart, J. S., 121, 122, 123, *162*, *164*
Harvey, E., 21, *164*
Hasselbalch, K., 8, 9, *163*
Haughton, T. M., 77, 78, *165*
Hays, F. A., 148, *165*
Hazelhoff, E. H., 22, 34, 96, 115, 116, *165*
Heller, J., 87, *165*
Hemingway, A., 104, *165*
Hemmingsen, E. A., 5, 54, 55, *165*
Hiller, G., 142, 152, *162*
Hilpert, P., 139, *168*
Hinton, H. E., 94, *165*
Holland, R. A. B., 13, *165*
Hopkinson, D. A. W., 104, *169*
Howson, M. G., 106, *164*
Hughes, G. M., 31, 34, 35, 37, 38, 39, 40, 41, 42, 43, 45, 48, 54, *165*, *168*
Hurtado, A., 124, 125, 126, 127, *165*
Hustead, R. F., 103, *163*
Hutchson, W. H., 80, *172*
Hyman, L. H., 23, *165*

I

Irving, L., 10, 133, 135, *165*, *170*
Ito, T., 87, *165*

J

Jakubowski, M., 54, *165*
Johansen, K., 10, 12, 28, 29, 30, 31, 33, 46, 58, 59, 60, 61, 62, 63, 64, 65, 66, 67, 68, 69, 70, 71, 72, 73, 74, 75, 77, 78, 79, 130, 134, 135, 136, *165*, *166*, *167*
Johnson, M. L., 79, *166*
Johnson, R. P., 103, *163*

K

Kanwisher, J. W., 88, 89, 90, 91, *166*
Kao, F., 105, *166*
Kawishor, J., 121, *170*
Kellogg, R. H., 124, *166*
Kendall, J. I., 49, *168*
Kerkut, G. A., 77, 78, *165*
Kjeldsen, K., 126, *168*
Kleerekoper, H., 41, *169*
Klika, E., 65, *166*
Knights, B., 31, 34, *165*
Kok, B., 27, *172*
Krogh, A., 2, 4, 8, 9, 21, 24, 45, 57, 80, 82, 83, 86, 97, 98, 107, 117, 155, *163*, *166*, *167*
Kruysse, A., 25, 36, 44, 46, 47, 55, *171*
Kuhl, D. K., 141, *169*
Kuhn, H. J., 158, 161, *167*
Kuhn, W., 158, 161, *164*, *167*
Kutchai, H., 152, 153, 154, *167*

L

Larimer, J. J., 109, *167*
Lasiewski, R. C., 121, *170*
Laverack, M. S., 77, *167*
LeFebvre, E., 122, *167*
Lelek, A., 65, *166*
Lenfant, C., 10, 12, 29, 30, 31, 33, 46, 60, 61, 64, 66, 67, 68, 69, 71, 72, 73, 74, 75, 129, 130, 134, 135, *166*, *167*
Levy, R. I., 88, *167*
Libby, W. F., 17, *162*
Lindroth, A., 34, *167*
Lokhorst, W., 146, *170*
Longmuir, I. S., 5, *167*
Longo, L. D., 141, 143, *167*, *169*
Løvø, A., 134, *162*

M

Maas, J. A., 97, *167*
McCobe, M., 5, *168*
McCutcheon, F. H., 76, *168*
MacDougall, J. D. B., 5, *168*
Malyusse, M., 55, *162*
Manwell, C., 27, 28, 142, *168*
Maren, T. H., 14, *168*
Marti, E., 161, *167*
Martin, A. W., 28, 29, 77, 78, 79, *166*
Mecklenburg, T. A., 61, *166*
Metcalfe, J., 139, 140, 141, 142, 143, 144, *162*, *168*
Mikhalev, Y., 54, *165*
Miller, P. L., 80, 83, 84, *168*

Mitchell, R. A., 129, 130, *168*
Moll, W., 5, 139, 140, 141, 142, 144, *168*
Mommaerts, W. F. H. M., 27, *172*
Morello, J. A., 14, *163*
Mossman, H. W., 139, *168*
Muir, B. S., 37, 49, *168*
Mulhausen, R., 126, *168*
Munday, K. A., 77, 78, *165*
Munshi, D., 61, *168*

N

Nemenz, H., 94, *168*
Nicloux, M., 52, 54, *168*
Novy, M. J., 142, 143, *168*

P

Paganelli, C. V., 92, 93, *168*, *169*
Parer, J. T., 113, 114, 139, 142, 143, *168*, *169*
Parry, G., 37, *169*
Pasztor, V. M., 41, *169*
Pattle, R. E., 103, 104, *169*
Pearson, O. P., 122, *169*
Pelseneer, P., 26, *169*
Perútz, M. F., 7, *169*
Petersen, A. J., 130, *167*
Petersen, J. A., 61, 66, 68, 69, 71, 72, 134, 135, *166*, *167*
Piiper, J., 46, *169*
Pilson, M. E., 31, *169*
Postma, N., 80, *163*
Power, G. G., 141, *169*
Precht, H., 97, *169*
Prosser, C. L., 21, 99, 100, 142, *169*
Purves, M. J., 105, *162*

R

Rahn, H., 92, 93, *168*, *169*, *172*
Rakestraw, N. W., 17, *169*
Ramel, A., 161, *167*
Ramos, J., 129, *167*
Randall, D. J., 55, *162*
Read, K. R. H., 29, *169*
Redmond, J. R., 31, 32, 33, 34, *169*
Reinhardt, W., 142, 152, *162*
Remmers, J. E., 106, 107, 108, 130, *171*
Remotti, E., 149, *169*
Reynafarje, C., 129, *167*
Richards, A. B., 87, *169*
Riess, J. A., 36, *169*
Riggs, A., 109, 111, *169*
Robinson, D., 133, *169*
Romanoff, A. J., 147, 148, 149, *170*
Romanoff, A. L., 147, 148, 149, *170*
Romijn, C., 146, 149, 150, *170*
Root, R. W., 9, 10, 48, *170*
Ross, J., 149, 150, *170*
Rossi-Fanelli, A., 28, *170*
Roughton, F. J. W., 13, 81, *163*, *170*
Roy, O. Z., 121, 122, 123, *162*, *164*
Ruud, J. T., 54, *170*

S

Salt, G. W., 115, 118, 119, *170*
Satchell, G. M., 72, *170*
Sato, T., 27, *170*
Saunders, R. L., 45, 46, 47, *170*
Savage, R. M., 76, *170*
Scammell, C. A., 31, 34, *165*
Schmidt-Nielsen, K., 61, 66, 68, 69, 71, 72, 107, 109, 110, 121, 130, 134, 135, *163*, *166*, *167*, *170*
Schneiderman, H. A., 88, *167*, *170*
Scholander, P. F., 5, 16, 133, 135, 137, 156, 157, 158, 160, *164*, *165*, *170*
Schulz, H., 61, *170*
Schumann, D., 46, *169*
Schwartz, J. B., 79, *163*
Schwarz, R. H., 141, 143, *167*
Sejrsen, P., 80, *171*
Semper, C., 57, 58, *171*
Servinghaus, J. W., 130, *171*
Smith, A. H., 126, *163*
Soloff, L. A., 103, *171*
Sørensen, S. C., 130, *171*
Staub, N. C., 100, *171*
Steen, J. B., 9, 10, 14, 15, 23, 24, 25, 26, 36, 37, 44, 46, 47, 48, 50, 53, 54, 55, 58, 145, 152, 153, 154, 156, 159, 161, *162*, *164*, *167*, *171*
Stickney, J. C., *171*
Strawinski, S., 75, *171*
Stride, G. O., 83, 94, *171*
Sturkie, P. D., 115, *171*
Sumbardo, A. H., 148, *165*
Sutnick, A. I., 103, *171*

T

Teal, J. M., 20, *163*
Tenney, S. M., 106, 107, 108, 130, *171*
Tierney, D. F., 103, *163*
Torrance, J., 129, *167*
Tuft, P. H., 87, *171*
Turitzin, S. N., 10, *171*

V

Valdivia, E., 126, *171*
Van Liere, E. J., *171*
Visschedijk, A. H. J., 150, 151, *171*

W

Wagner, H. N. Jr., 141, *169*
Wallengren, H., 57, *171*
Wangenstein, O. D., *172*
Waterman, T. H., 31, *172*
Weibel, E. R., 100, 101, 102, *172*
Weis-Fogh, T., 81, 83, 84, 85, *172*
White, P. T., 143, 151, *172*
Whitford, W. G., 80, *172*
Wigglesworth, V. B., 82, 83, *172*
Williams, C. H., 88, *170*
Willmer, E. N., 49, 52, 53, *172*
Wilson, D., *172*
Winterstein, H., 57, *172*
Wolvekamp, H. P., 27, 31, *172*

Y

Yonge, C. N., 26, *172*
Yu, C. I., 11, 130, *162*

Z

Zeuthen, E., 115, 116, 117, 118, 119, 120, 121, *170*, *172*

SUBJECT INDEX

A

Acetylcholine, 55
Acidosis, 58, 134, 138
Active transport, 2, 37, 82
Activity and respiratory function, 29, 48, 83, 99, 107
Adaptations
  of air breathing fish, 58
  of aquatic insects, 92
  of diving animals, 132
  to high altitude, 124
  to hypoxic water, 32, 49, 57
Adrenaline, 55
Aeshnia, 84
Aestivation, 64
Air
  as respiratory medium, 16, 19, 56
  atmospheric, 1, 16
  capillaries, 114, 116
  sacs, 102, 114, 118, 120
Allantois, 149
Alpha Helix Expedition, 18, 49, 57, 132
Alveolar
  capillaries, 102
  gas, 2, 19, 98, 100
  volume, 108, 124
Alveoli, 65, 100
  stability of, 102
Amazon river, 18, 49, 57, 132
Amia calva, 64, 73
Amphibians
  blood respiratory properties, 70, 99
  cutaneous respiration, 80
  lungs, 65, 100, 104
  metamorphosis, 75
Anabas testudineus, 61
Anaerobic metabolism, 136, 138
Ancula, 57
Anguilla vulgaris, 58, 75, *see also* Eels
Annelids
  aquatic, 25
  terrestial, 77
Antarctic fish, 54
Aortic body, 105
Aphelocheirus, 92
Aplysia deplians, 28
Arachnoids, 95
Area inflata, 27
Arion, 97

B

Bagrus, 49
Birds,
  at high altitudes, 130
  blood respiratory properties, 99
  diving, 135
  eggs, 142, 146
  lungs, 114, 150
    gas distribution, 118
    gas exchange, 116
    ventilation, 19, 117, 150
Birgus latro, 57
Blaberus, 83
Blatella, 83
Blood
  carbon dioxide tension, 2, 9, 73, 105
  carbon dioxide transport, 11
  cells, *see* Erythrocytes
  circulation, *see* Circulatory systems
  gases, reaction rate, 13
  oxygen tension, 2, 7, 105
  pH value of, 10, 105
  respiratory properties, *see* Oxygen affinity, Oxygen capacity
Body size and lung function, 106
Bohr-effect (Bohr-shift), 9
  in annelids, 77
  in crustaceans, 32
  in diving animals, 133
  in fish, 52
  in lungfish, 73
  in mammals, 100, 109, 145
  in molluscs, 26, 31
  of foetal blood, 145
  temperature sensitivity, 15
Bombina, 100
Bradycardia, 135
Bucco-pharyngeal gas exchange, 58, 132
Bullhead, 45, 48
Byrsotria, 83

C

Callionymus, 40
Caloppa granulata, 34
Capillaries
  air, 114, 116
  allantoic, 149
  alveolar, 102
  placental, 139
  swimbladder, 155
Carassius, *see* Goldfish
Carbon dioxide
  and oxygen affinity, 9
  atmospheric, 1, 16
  cyclic release, 87
  equilibrium curves, 13
  reaction with blood, 13
  solubility, 2, 20
  spiracle control by, 96
  tension, 2, 9, 73, 105
  transport of, 11
Carbonic anhydrase, 14, 17, 109
Carcinus, 57
Carcinus maenas, 34
Cardita floridana, 28
Carotid body, 105
Carp, 45, 52, 54
Cat, 109
Cecropia, 88
Cephalopods, 11, 26, 29, 79
Chaenichthys rugosus, 54
Chaenocephalus aceratus, 54
Chamsocephalus esox, 54
Chelys fimbriata, 134
Chicken, 116, 147
Chyrptochiton stelleri, 27
Circulatory systems
  and respiratory efficiency, 6
  in annelids, 79
  in crustaceans, 31
  in echinoderms, 22
  in fish, 36, 46, 55, 61
  in lungfish, 66
  in mammals, 100
  in molluscs, 25, 29
  of embryos, 149
  of foetus, 139
Cladocera, 32
Cloacal respiration, 132
Coelenterates, 22
Common sucker, 52
Con-current respiratory flow, 6
Cossus larvae, 82
Counter-current blood flow
  in placenta, 139, 145
  in swimbladder, 155
Counter-current respiratory flow, 6, 19
  in crustaceans, 34
  in fish, 46
  in molluscs, 25, 29
Crow, 115, 116
Crustaceans
  aquatic, 31
  intertidal, 57
  Root-effect in, 11
  terrestial, 98
Ctenidia, *see* Gills of molluscs
Ctenophores, 22
Cutaneous gas exchange, 54, 58, 77, 80, 132
Cyclic carbon dioxide release, 87
Cyprinus, *see* Carp

D

Dana Expedition, 18
Daphnia, 32
Dead space, 106
Decompression sickness, 132, 135
Desmognathus quadramaculatus, 80
Diapause, 89
Diffusion
  barrier, 6
  capacity, 5
  coefficient, 3, 5, 21
  constant, 24, 45
  in air capillaries, 116
  in alveoli, 104
  in trachae, 82
  process of, 1
Diphosphoglycerate (2,3-DPG), 11, 130
Dipnoi, *see* Lungfish
Diving animals, 132
Dog, 86, 141
Dragon-fly, 80, 84
Drosophila, 85
Dual breathers, 57
Duck, 121, 134, 138

E

Earthworm, 77
Echinoderms, 22
Eel, 15, 44, 46, 48, 52, 54, 57, 63, 75, 161
Efficiency of respiratory mechanism, 6, 98, 107, 112
Eggs, 21
  amphibian, 75
  birds, 146
  insects, 87, 95
Electrophorus electricus, 60, 75
Embryos, 21, 146
Ephydra larvae, 94

Erythrocytes
  of fish, 10, 52
  of higher vertebrates, 99
  of molluscs, 27
  oxygenation of, 13
Esox, see Pike
Exchange ratio, 73

F

Fick's Law, 117
Fish
  activity and respiration, 48
  air breathing, 58
  gill apparatus
    adaptations of, 48, 55
    gas exchange in, 42
    structure, 34
    ventilation, 37
  haemoglobin, 48, 54, 142
  swimbladder, 2, 11, 63, 155
Flow patterns, 6, *see also* Counter-current respiratory flow
Foetus, 139

G

Gadus pollachius, 37
Gas
  concentration, 2, 160
  exchange
    aquatic, 21
    aerial, 77
    bimodal, 56
    estimation of, 3, 21
    facilitation of, 5
  tension, 2
Gills
  adaptations of, 48, 58
  ion exchange in, 20, 37, 55
  of crustaceans, 31
  of fish, 34, 65
  of molluscs, 25
  reduction in, 60, 66
  resistance of, 40
  surface area of, 26, 31, 37, 48
Glossoscolex giganteus, 77
Gobius, 41
Goldfish, 54, 57
Grapsus, 57
Guinea-pig, 130

H

Haematocrit, 99, 126
Haemocyanin (Hcy), 7
  of crustaceans, 32
  of molluscs, 26, 31
  oxygen dissociation curve, 26
Haemoglobin (Hb)
  abnormalities in, 113
  at high altitudes, 130
  carbon dioxide transport by, 12
  facilitation by, 5
  foetal, 142
  molecular structure, 7
  of annelids, 77
  of crustaceans, 32
  of fish, 48, 54, 142, 155
  of higher vertebrates, 98
  oxygen equilibrium curves, 8
Hairpin counter-current multiplication principle, 158
Haldane-effect, 13
Haliotes corrugata, 31
Heart, 29, 47, 66, 69, 71, 135
Heat capacity of air and water, 20
Helicostoma, 80
Helix pomatia, 97
High altitude, effect of, 124
Hole fraction, 85
Holothuria tubulosa, 22
Homarus americanus, 32, 33
Homeothermy, 20, 71, 99, 104, 121
Hydrogen ion concentration, *see* pH
Hydrostatic organs
  in fish, 63, 155
  in molluscs, 97
Hyperventilation, 104, 130
Hypoxic environments, 32, 49, 57, 69, 124, 132

I

Icefish, 54
Iguana iguana, 135
Insects
  aquatic, 92
  cyclic carbon dioxide release, 87
  plastron breathing, 92
  terrestial, 80
  tracheal system, 80
  ventilation, 82, 88
  water conservation, 87
Intestinal air breathing fish, 63
Ion exchange, 20, 37, 55

L

Lactic acid, 11, 58, 136
Lamellae
  of fish, 36, 59
  of molluscs, 25
Lamellar gills of crustaceans, 31

Lates, 49
Lepidosiren paradoxa, 66
Leuciscus erythrophthalamus, 57
Libellula, 84
Locust, 80, 84
Loxorhynchus gradis, 33
Lumbricus, *see* Earthworm
Lungfish
  blood respiratory properties, 73
  heart, 71
  lungs, 63, 64, 104
  ventilation, 72
Lungs
  air capillary, 114
  alveolar, 100
  diffusion, 95
  fluid lining of, 102
  function and body size, 106
  of birds, 114, 150
  of lungfish, 64
  of mammals, 100
  of molluscs, 96
  ventilation of, 19, 72, 97, 98, 104, 114, 117, 130, 138
  volume, 118, 134
Lymnea, 80

M

Mackerel, 37, 40, 48
Mammals
  at high altitudes, 124
  blood respiratory properties, 99
  body size relationships, 105
  diving, 133, 136
  lungs, 100
  placental gas exchange, 139
Metamorphosis, 76
Molluscs
  aquatic, 25
  intertidal, 57, 80
  terrestial, 95
Mosquito larvae, 82
Mouse, 107
Mouth breathing fish, 58
Myoglobin, 11, 27, 133

N

Nematodes, *see* Roundworms
Neoceratodus forsteri, 66
Nitrogen, 1, 16, 17, 90, 92
Notonecta, 93
Nyctobra, 83

O

Octopus dofleini, 29
Osmosis and tracheal systems, 82
Ostrich, 121
Oxycera larvae, 94
Oxygen
  affinity of blood, 9, 11
    at high altitudes, 126, 130
    carbon dioxide tension and, 9
    of amphibia, 73, 76
    of annelids, 77
    of birds, 133, 142, 152
    of crustaceans, 32
    of fish, 48, 73
    of mammals, 109, 113, 126, 130, 133, 142
  arterio-venous differences in, 47, 66, 69, 99
  atmospheric, 1, 16
  capacity of blood, 7
  concentration by rete, 158
  consumption in insects, 83, 89
  content, 6, 7
  debt in eels, 58
  dissociation curves, 8, 26, 100, 113, 125, 152
  equilibrium curves, 8, 31, 52, 66, 73, 77, 127, 134, 144, 161
  extraction, 6
    in aquatic animals, 22, 23, 29, 34, 46, 134, 157
    in terrestial animals, 98
  reaction rate with blood, 13
  solubility, 2, 16
  stores, 132
  tension, 2, 7, 105
  transport, 7
  uptake, 2, 6
    and body size, 107
    and ventilation volume, 104
    measurement of, 24, 44

P

Pachygrapsus crassipes, 33
Paku, 52
Palinurus, 32
Panulirus interruptus, 33
Paramecium, 21
Paratrygon sp., 57
Parazoa, *see* Sponges
Partial pressure, 1
Partial pressure gradients
  across respiratory membranes, 2, 4, 6, 11, 19, 44, 98, 107, 124

Partial pressure gradients (*cont.*)
between blood and tissue, 6, 11, 31, 86, 135
Periophthalmus schlosseri, 63
Peripatus, 81
Periplaneta, 83
Permeability coefficient, 4
pH of blood
and oxygen affinity, 9 *see also* Bohr-effect
and oxygen capacity, 9 *see also* Root-effect
and respiratory regulation, 105
measurement of, 10
Pigeon, 121
Pike, 54
Pillar cells, 37
Placental gas exchange, 139
Planaria, 21, 22
Planorbis corneus, 28
Plastron breathing, 92
Podia of sea urchins, 23
Podocnemys sp., 132
Poiseuilles's equation, 39
Porcellio, 98
Proteus, 65, 100
Protopterus aethipicus, 66, 104
Protozoa, 21

R

Rabbit, 115, 139, 141
Rana catesbeiana, 13, 76
Rana esculenta, 75, 100
Rana olympicus, 100
Rana pipiens, 70
Rana temporaria, 76
Reptiles
blood respiratory properties, 72, 99
lungs, 72, 100, 114
respiratory adaptations, 132, 134, 138
Respiratory
control centre, 105
media, 16
membranes
permeability, 2, 3
surface area of, 3 *see also* Gills, Lungs
thickness of, 3, 24, 31, 37, 49, 61, 65
organs, *see* Gills, Lungs
organs, accessory, 56, 132
pigments, 7, 26, *see also* Haemocyanin, Haemoglobin
quotient (RQ), 73, 89
Rete mirabile, 155
Rhodnius, 87
Root-effect (Root-shift), 9, 15, 52, 73, 160
Rotatoria, 21
Roundworms, 22

S

Salt-secreting cells, 20, 37, 55
Scaphiopus couchi, 100
Schistocera gregaria, 84
Scomber, *see* Mackerel
Scorpacnichtys, 142
Scorpion, 96
Scutigera, 98
Seals, 135, 136, 138
Sepia officinalis, 27
Sheep, 130, 139
Solubility coefficient, 2
Spider, 95
Sponges, 22
Squalus suckleyi, 142
Sternothoerus merior, 138
Strongylocentrotus droebachiensis, 24
Sucker fish, 45, 52
Swimbladder, 2, 11, 63, 155
Symbranchus marmoratus, 58, 63
Sympetrum, 84

T

Tadpoles, 75
Tarantula, 95
Tench, 40
Toadfish, 48
Trachea, 106, 111, 115
Tracheal system of insects, 80
Trichogaster trichopterus, 61
Tuna fish, 20
Turtles, 100, 132, 134

V

Ventilation
and cardiac coordination, 29, 66, 70, 104
and respiratory efficiency, 6, 98
at high altitudes, 130
of gills, 25, 29, 34, 37, 61, 64
of lungs, 19, 72, 97, 98, 104, 114, 117, 130, 138
of tracheal system, 82
regulation of, 98, 104, 130
resistance, 106, 112
tidal, 19
Vesicular gills of crustaceans, 31

W

Water
as respiratory medium, 16, 21, 92, 132
conservation of, 87, 97, 152
gaseous components, 16
hypoxic, 32, 49, 57
loss, 122
Whales, 135